PRESERVING OUR PLANET

APPLYING GIS

PRESERVING OUR PLANET

GIS FOR CONSERVATION

Edited by
David Gadsden
Matt Artz

Esri Press
REDLANDS | CALIFORNIA

Esri Press, 380 New York Street, Redlands, California 92373-8100
Copyright © 2022 Esri
All rights reserved.
Printed in the United States of America.

ISBN: 9781589487215
Library of Congress Control Number: 2022943529

The information contained in this document is the exclusive property of Esri or its licensors. This work is protected under United States copyright law and other international copyright treaties and conventions. No part of this work may be reproduced or transmitted in any form or by any means, electronic or mechanical, including photocopying and recording, or by any information storage or retrieval system, except as expressly permitted in writing by Esri. All requests should be sent to Attention: Director, Contracts and Legal Department, Esri, 380 New York Street, Redlands, California 92373-8100, USA.

The information contained in this document is subject to change without notice.

US Government Restricted/Limited Rights: Any software, documentation, and/or data delivered hereunder is subject to the terms of the License Agreement. The commercial license rights in the License Agreement strictly govern Licensee's use, reproduction, or disclosure of the software, data, and documentation. In no event shall the US Government acquire greater than RESTRICTED/LIMITED RIGHTS. At a minimum, use, duplication, or disclosure by the US Government is subject to restrictions as set forth in FAR §52.227-14 Alternates I, II, and III (DEC 2007); FAR §52.227-19(b) (DEC 2007) and/or FAR §12.211/12.212 (Commercial Technical Data/Computer Software); and DFARS §252.227-7015 (DEC 2011) (Technical Data–Commercial Items) and/or DFARS §227.7202 (Commercial Computer Software and Commercial Computer Software Documentation), as applicable. Contractor/Manufacturer is Esri, 380 New York Street, Redlands, California 92373-8100, USA.

Esri products or services referenced in this publication are trademarks, service marks, or registered marks of Esri in the United States, the European Community, or certain other jurisdictions. To learn more about Esri marks, go to: links.esri.com/EsriProductNamingGuide. Other companies and products or services mentioned herein may be trademarks, service marks, or registered marks of their respective mark owners.

For purchasing and distribution options (both domestic and international), please visit esripress.esri.com.

On the cover: Photograph by David Gadsden.

CONTENTS

Introduction — vii
How to use this book — ix

PART 1: CONSERVATION LAND MANAGEMENT — 1

Managing protected areas — 3
Esri and the National Geographic Society

Empowering indigenous communities with conservation solutions — 10
African People & Wildlife

Enabling self-service access to visualize data — 20
Open Space Institute

Using smart maps to save lions and solve human–wildlife conflict — 26
Born Free Foundation

Wielding location intelligence to fight poachers — 32
Chengeta Wildlife

Hollywood tricks and location technology catch poachers — 40
Durrell Institute of Conservation and Ecology at the University of Kent

Where the bats go — 47
Bat Conservation International

Finding balance: Helping people and predators coexist — 55
Tanzania National Parks

PART 2: LANDSCAPE CONSERVATION — 63

Mapping native forests in western Canada — 65
Conservation North

Seeing mule deer decline from above and connecting the dots — 72
Nevada Department of Wildlife

Mapping America's land and sea: A time for precision conservation — 81
Esri

Maps cut through the fog to help preserve unique ecosystems — 88
Servicio Nacional de Áreas Naturales Protegidas por el Estado

Web map brings together conservation and green energy development — 97
The Nature Conservancy

Scientists collaborate to map biodiversity and the human footprint — 103
Alberta Biodiversity Monitoring Institute

Conserving a network of climate-resilient lands — 110
The Nature Conservancy

NEXT STEPS — 117

Contributors — 121

INTRODUCTION

CONSERVATION ORGANIZATIONS FACE INCREASINGLY multifaceted challenges to preserving the biodiversity of habitats in our rapidly changing world. Modern conservation technologies enable efficient, real-time observations to monitor natural areas, leading to new insights and understanding through conservation science. Grounded in geography, conservation GIS enables improved understanding of the complex web of threats, opportunities, and challenges facing our natural world.

Geospatial infrastructure for conservation

With more than 50 years of software development, GIS delivers a comprehensive toolset for conservation professionals. Advanced conservation science reveals insights into complex environmental challenges to improve our understanding of the interdependencies of native species and human activity and informs strategies to mitigate threats presented by climate change and unmanaged development. GIS equips conservation professionals with maps and apps optimized for common conservation workflows, underpinned by a vast collection of analysis-ready, open geospatial data to support and improve local and global biodiversity conservation efforts.

Conservation technologies

GIS provides a comprehensive set of technology solutions for efficient, science-backed conservation programs.

- **Observe:** GIS offers integrated mobile apps for planning, conducting, and analyzing field observations with integrated management tools to take your mobile work to scale.

- **Organize:** GIS allows you to manage vast datasets in a secure cloud or hybrid data stores to power real-time dashboards of conservation operations, enhanced with content from ArcGIS® Living Atlas of the World.

- **Analyze:** GIS helps you discover new conservation dynamics through modeling the natural world with the most comprehensive and open set of analytical methods and spatial algorithms available.

- **Collaborate:** GIS allows you to accelerate effective partnerships and public engagement with shared maps, apps, and stories to inform broad and transparent conservation initiatives.

Conservation strategies

Preserving Our Planet: GIS for Conservation illustrates how organizations apply The Geographic Approach—the use of geography to solve problems and make decisions—to land management and conservation. The stories in this book show readers how they can use GIS to gain a geographic perspective and integrate spatial reasoning into conservation activities. The book presents location intelligence as another layer of knowledge that organizations can add to their experience and expertise and incorporate into their daily operations and planning. The book concludes with a section about getting started with GIS, which provides strategies, tools, and suggested actions for incorporating location-based conservation intelligence into decision-making and operational workflows.

HOW TO USE THIS BOOK

THIS BOOK IS DESIGNED TO HELP DRAW ATTENTION TO THE issues facing the global conservation community right now. It is a guide for taking first steps with GIS and applying location intelligence to decisions and operational processes to solve common conservation problems. You can use this book to identify where maps, spatial analysis, and GIS-powered apps might be helpful in your work and then, as next steps, learn more about resources to get started.

Learn about additional GIS resources for natural resources by visiting the web page for this book:

go.esri.com/pop-resources

PART 1
CONSERVATION LAND MANAGEMENT

THE EXPANDING HUMAN FOOTPRINT AND THE GROWING threat of climate change place immense pressure on biodiversity conservation and long-term management plans. As global biodiversity continues to decline, conservationists must effectively manage and safeguard designated protected areas. GIS conservation solutions are designed through extensive engagement with leading protected area management organizations. These solutions strengthen operational capabilities, reduce training and implementation costs, and support informed and timely conservation land management decisions.

Conservation solutions for protected areas

Successfully managing a protected area is a difficult challenge. Disease, fire, and climate-related disruptions within protected areas coupled with external population and development pressures require effective management systems to support conservation managers. ArcGIS Solutions is a system within ArcGIS Online that offers a collection of apps, dashboards, and analysis and reporting web applications to support conservation management. Deploying the solutions in ArcGIS Online is automated so that managers of protected areas can focus on configuring an existing system to meet their needs instead of implementing a system from scratch. These solutions, now in use globally, offer these core capabilities:

Track illegal activity: Many protected areas face pressure from illegal activities including poaching, encroachment, and deforestation. Conservation solutions streamline the collection of field observations and provide a digital chain of custody for law enforcement actions, informing real-time monitoring dashboards to support timely decision-making.

Monitor wildlife: Location-enabled wildlife field observations help managers of protected areas evaluate the effectiveness of management actions. They can monitor trends derived from field information through spatial analysis to inform and evaluate management strategies for protected areas. Conservation solutions offer a data-driven approach to wildlife management in protected areas.

Mitigate wildlife conflicts: The job of managing protected areas occurs within a tapestry of wild and human-populated landscapes. Historic encroachment into natural areas and corridors and, in rare cases, recovery of predator populations have increased conflicts between humans and wildlife in local communities. Conservation solutions support the collection and monitoring of incidents adjacent to a protected area and support planning and actions to reduce conflicts.

Maintain park infrastructure: Conservation land management involves protecting natural areas while providing basic infrastructure for visitors and recreation opportunities. Conservation solutions support planning of infrastructure projects and maintenance programs and efficient management of staff and resources in protected areas.

MANAGING PROTECTED AREAS

Esri and the National Geographic Society

GLOBAL BIODIVERSITY FACES UNPRECEDENTED THREATS from habitat loss, poaching, and climate change. These factors increase the fragmentation of critical habitats and threaten key species. Integrated, technology-based systems for managing and protecting natural areas are desperately needed.

Building upon decades of collaboration, Esri® and the National Geographic Society have formed a strategic partnership to provide advanced solutions for managing protected areas. Through the alignment of Esri's geospatial technology and National Geographic's expertise, the organizations are committed to advancing partnerships, technologies, capacity, and methods for management and storytelling to preserve and sustain natural areas.

ArcGIS Solutions includes a collection of apps developed by Esri and National Geographic specifically for managing protected areas. These apps provided integrated tools that can be configured to serve the needs of conservation area managers around the world.

This suite of tools serves many roles in conservation, as illustrated in a sampling of comments from conservation managers:

- "I'm a community outreach specialist. I need to connect local villages to park management and ensure that people benefit economically from the tourism the park attracts."

- "I manage my park's facilities. I need to keep track of the park's infrastructure, so everything runs smoothly."

- "I'm a wildlife biologist. I want to help ensure that animal populations remain stable, and I want freedom to do my research."

- "I'm the park superintendent. I need to keep things running, connect to local communities, and ensure that the habitats and animal populations in my park remain healthy."
- "I lead an international NGO. I want to conserve lands and species, but I know I can't do that without connecting to local stakeholders and understanding their needs."
- "I'm a game warden. I need to know where the poaching threats are greatest and where the wildlife is."

ArcGIS Solutions for Protected Area Management can meet the needs of all these stakeholders and enable land managers to be more effective in their work.

A comprehensive toolset

Today, web-based mapping tools are revolutionizing the way conservation managers create, manage, and disseminate information needed for decision-making. The solutions provide managers of protected areas with configurable technology for managing and protecting parks and sensitive conservation areas.

The Ecology, Infrastructure, Outreach, and Protection solutions address the diverse challenges of managing parks by providing a secure and flexible framework for deploying useful applications and integrating other conservation technology to support key roles and workflows in park operations.

Ecology solution

The Ecology solution is designed to enable wildlife management activities such as collecting species observations in the field and analyzing and reporting wildlife trends. For example, a park ecologist who manages wildlife species occurrence data can use the tool to monitor species populations, health, fecundity, and range.

Infrastructure solution

The Infrastructure solution is designed to enable asset management activities, including collection of asset information in the field, managing of maintenance crews, monitoring productivity in near real time, and visualizing the location and condition of assets. A maintenance staff member, for example, can use this tool to collect and update asset information and then update managers on the condition of these assets.

Outreach solution

The Outreach solution is designed to enable community engagement activities, including collection of human and wildlife conflict information in the field, monitoring of those incidents in near real time, and communication with protected area stakeholders and visitors about the resources and activities under management. For example, an outreach coordinator can use a map that identifies these resources and activities to support management plans as they relate to stakeholders and constituents in and around the park.

Protection solution

The Protection solution is designed to enable law enforcement activities, including collecting incident information in the field, managing patrol mandates, monitoring active patrols and incidents in near real time, and analyzing and reporting on incident trends. A park ranger, for example, can use this tool to collect, maintain, and submit protection incidents so that protection managers can see the status of incidents and plan patrols and enforcement actions.

Solutions in action

Many organizations use ArcGIS Solutions for Protected Area Management in their daily work. Let's take a closer look at how some leading African organizations use these apps and tools.

African Parks

African Parks is a nonprofit conservation organization that assumes responsibility for the rehabilitation and long-term management of 19 national parks and protected areas in 11 countries across Africa, in partnership with governments and local communities.

To effectively manage, protect, and restore parks across more than 14.8 million hectares (almost 36.6 million acres), the organization and its partners implement a range of core functions from law enforcement and biodiversity conservation to community and enterprise development and infrastructure. They integrate innovative technological solutions to improve monitoring and overall management of these protected areas.

African Parks uses ArcGIS Solutions for Protected Area Management to monitor and enhance park management workflows. For instance, it can manage park roads, buildings, and park boundaries using up-to-date map data.

GIS Rangers compiling field-collected data. Photo by Mia Collis.

The parks require customized solutions and workflows to meet the needs of diverse landscapes. Park staff use field apps and mapping software to save time and improve outcomes for people and wildlife. They also use ArcGIS Dashboards to monitor sensitive species such as rhinos, elephants, and giraffes. These systems closely align with law enforcement efforts to monitor wildlife populations in parks, boosting security so that people and wildlife can thrive.

Peace Parks Foundation

The Peace Parks Foundation aspires "to reconnect Africa's wild spaces to create a future for humans in harmony with nature."

The foundation works to renew and preserve ecosystems that span international boundaries by establishing conservation areas. In doing so, the foundation strives to protect biological diversity, natural resources, and cultural heritage. The foundation also supports the goal of reducing poverty through the economic benefits of ecotourism.

Peace Parks requires tools to understand the needs of communities living in and adjacent to protected areas. It uses this knowledge to develop long-term solutions. Mapping technology supports the foundation in its management decisions, tourism planning, and fund development.

ArcGIS Pro provides analysis tools and helps managers as they make decisions in protected areas. Field apps such as ArcGIS Collector and ArcGIS Survey123 help provide information about communities living in and adjacent to conservation areas.

Conservation-focused GIS training on ArcMap™ and field data collection are being used in partnership with local universities in protected areas. These trainings ensure that staff are up to date with the latest tools and best practices in mapping technology.

Peace Parks Foundation GIS rangers using ArcGIS in the field.

The Jane Goodall Institute

Dr. Jane Goodall has long supported local communities in her scientific and conservation work with chimpanzees in the United Republic of Tanzania. Her work has evolved into the Lake Tanganyika Catchment Reforestation and Education program (Tacare), the Jane Goodall Institute's approach to put the needs of local people at the forefront of conservation activities.

As Tacare has evolved, the institute has increasingly used geospatial technologies to integrate traditional knowledge with the best GIS data and high-resolution satellite imagery available to guide and inform community decision-making and conservation strategies. The institute has developed GIS layers to understand chimpanzee behavior, movements, and habitat use, as well as to model habitat suitability and human land-use change. It has paired this data with field mobile data collection apps to enable communities to monitor and enforce plans for their village forest reserves and land use.

Jane Goodall's Tacare program works closely with local communities so that they can decide on the best approaches for conservation.

The approach of the Jane Goodall Institute and the mapping technologies it uses have supported the restoration and protection of chimpanzee habitats in western Tanzania, western Uganda, the eastern Democratic Republic of the Congo, and across a larger chimpanzee range as the organization seeks to scale the Tacare approach.

The Jane Goodall Institute supports local people and communities as they become stewards of their own natural resource management, governance, and sustainable livelihoods.

A version of this story titled "Managing Protected Areas: Innovative Solutions to Meet the Needs of Conservation Area Managers Around the World" appeared in a storytelling map on March 8, 2022.

EMPOWERING INDIGENOUS COMMUNITIES WITH CONSERVATION SOLUTIONS

African People & Wildlife

AFRICAN PEOPLE & WILDLIFE (APW) ENVISIONS A WORLD where Africa's people and wildlife coexist and thrive in vibrant, healthy landscapes. The team at APW works to realize this vision through innovative efforts in Tanzania and other African countries. These efforts, combined with community engagement expertise, drive effective, measurable, and lasting outcomes for communities and nature.

Elizabeth Naro, monitoring and evaluation and GIS lead for Tanzania People & Wildlife, has shared several powerful examples of efforts to find a balance between tradition and technology to foster innovation in conservation science.

APW is based in northern Tanzania, with its headquarters, the Noloholo Environmental Center, located at the southeastern border of Tarangire National Park. The organization was cofounded in 2005 by Dr. Laly Lichtenfeld and Charles Trout to turn negative trends in wildlife conservation around and understand environmental issues from the local point of view.

Home to some of Africa's most iconic protected areas, northern Tanzania is a biodiversity hot spot. But the country's protected areas are not fenced, which allows wildlife to roam freely along natural migratory routes that cross through human-dominated landscapes. This means that 92 percent of the rangeland is shared by people, livestock, and wildlife.

Naturally, this freedom of movement increases the likelihood of human–wildlife conflict. APW focuses its work on land outside formally protected areas, where finding the balance between

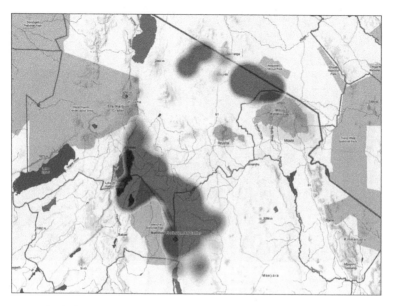

Areas shaded in brown indicate APW's big cat observation range for such wildlife as lions, cheetahs, and leopards in northern Tanzania.

human well-being and healthy ecosystems is critical for long-term sustainability.

Reducing human–wildlife conflict is a primary goal of APW's, and the use of geospatial technology through ArcGIS has furthered the organization's work in three main program areas: the Human–Wildlife Conflict Prevention Program, the Natural Resource Stewardship Program, and the Conservation Enterprise Program.

Human–Wildlife Conflict Prevention Program

When APW was founded in 2005—and for the first decade of its work—team members collected data on human–wildlife conflict incidents through paper forms. The forms were filled out by local Maasai warriors, whose traditional role in a community is to protect people and livestock from predators.

Constructing a living wall with *Commiphora* trees and chain-link fencing.

Warriors for Wildlife, an APW team that responds to human–wildlife conflict events across 30 communities in northern Tanzania, has provided invaluable data on conflict trends. The team also helped implement the Living Walls Program, in which *Commiphora* trees grow to form the posts of cattle corrals. These nature-friendly corrals are reinforced with chain-link fencing to create an impenetrable barrier.

In 2016, APW shifted to a mobile data collection system. However, analytics and visualizations were developed manually in Excel and other visualization and reporting solutions. In 2019, APW connected with the Esri conservation team and was given access to ArcGIS Solutions to help the organization manage protected areas.

"This really revolutionized the way we collect data in the field and provided a platform for automated output analysis, giving us more time to conduct outcome- and impact-level program evaluations that require more qualitative data," said Naro.

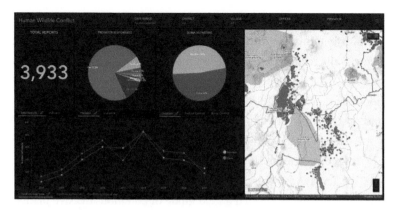

The Human–Wildlife Conflict dashboard shows total reports of conflicts and conflict trends over time.

One of the tangible outcomes of deploying ArcGIS Solutions is simple analytics in real time. ArcGIS Dashboards linked to live Survey123 data provides a current operational view of conflict data. The dashboards are configured to visualize the most appropriate information to support APW and its partner communities' understanding and management of conflict dynamics.

For instance, the percentage of conflict incidents by predator type can be quickly visualized revealing that hyenas (or *fisi* in Kiswahili) are responsible for over 70 percent of attacks on livestock. Looking deeper into the hyena attack charts reveals that 60 percent of these conflicts occur at the *boma*, or homestead, not in the shared rangeland.

Reviewing such information by political district allows APW to set priorities for living walls and other programs to target resources toward interventions most likely to make a positive impact. To visualize conflict data within the districts, heat maps are dynamically generated to further focus intervention efforts.

Working with communities and other local partners, APW has

planted more than 210,000 trees to build more than 1,500 living walls. By eliminating the need to cut trees to build traditional corrals, living walls revitalize habitats, improve landscape connectivity, and increase local climate change resilience. They also secure the livelihoods of more than 18,000 people living in rural northern Tanzania and protect a population of about 500 lions from retaliatory killing.

As the rangeland monitoring program developed, the village grazing committees expressed their need for more data on invasive species trends. Working with rangeland monitors across northern Tanzania, APW created a field guide of common invasive plants. The locally relevant names of each plant were then incorporated into rangeland monitoring Survey123 forms.

As the communities' traditional and observational knowledge was augmented with the incoming monitoring data, trends became apparent, such as the dangerous expansion of the invasive plant *Solanum incanum*, a species of nightshade.

After successfully transitioning the Human–Wildlife Conflict Prevention Program to ArcGIS, APW explored whether the same tools could be useful for rangeland management by community decision-makers.

Naro and her team transitioned the rangeland monitoring protocols to Survey123. By observing the grazing quality of critical pastures each month, pastoral communities are better able to manage rangeland to ensure that grass resources are sustained throughout the dry season.

The monitoring program involves working with the grazing committees in each village to select plots for monthly surveilling. Two local pastoralists from each community regularly measure grass height, vegetation cover, and a few other pasture quality metrics every 5 meters along a 100-meter transect, creating 20 samples.

This sampling data is then simply visualized in ArcGIS to help the grazing committees make informed management decisions.

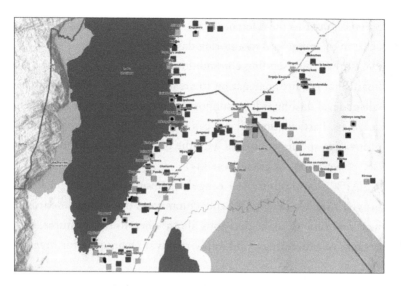

Pasture-monitoring plots along the Tarangire-Manyara corridor.

As the rangeland monitoring program developed, the village grazing committees expressed their need for more data on invasive species trends. Working with rangeland monitors across northern Tanzania, APW created a field guide of common invasive plants. The locally relevant names of each plant were then incorporated into rangeland monitoring Survey123 forms.

As the communities' traditional and observational knowledge was augmented with the incoming monitoring data, trends became apparent, such as the dangerous expansion of the invasive nightshade species.

Incorporating a related table in the Survey123 form simplifies the rangeland monitors' collection of invasive species present in the 20 samples along each transect for every plot.

APW's GIS team overcame the challenge of displaying multiple species in one chart by using an ArcGIS Arcade expression that sorted, filtered, and calculated the frequency of each species and populated that frequency back into the parent dataset. Linking the parent dataset to the reference feature layer powers a multiple-series chart in the rangeland monitoring dashboards.

Rather than sending communities to a complex regional dashboard that requires navigation to areas of interest, APW configures village-level dashboards to support local conservation management decisions. Participating villages can easily visualize how the presence of invasive species compares with the grass height and the pastoralists' observations of grazing quality.

APW's community-led rangeland monitoring approach has proved valuable. Village grazing committees use the dashboards in near real time to make decisions about when to rest pastures, shift grazing to other areas, and apply removal techniques for invasive species.

Another example of APW working to support sustainable livelihoods for rural communities is the Women's Beekeeping Initiative. Women's groups in communities near Tarangire National Park harvest honey to sell in local markets.

Since trees cannot be cut down in Tanzania if they support local enterprises, the beehives provide some protection for shade trees and pastures. However, the hives may also help restore larger portions of rangeland, since honeybees will travel about five kilometers from their hive for nectar and are more likely to select native species as fodder.

Using ArcGIS Pro, Naro's GIS team generated a five-kilometer buffer around each hive tree to show where the bees may pollinate native species. The result of this simple analysis is an estimated 439,000 acres of pasture that may benefit from the presence of the hives in the beekeeping program.

Naro hopes that ongoing work will incorporate the rangeland monitoring plot data to compare frequency trends for invasive species between plots within the beehives' zones of influence and those

Strategically placed beehives are used as a restoration technique.

outside. This integrated approach would help estimate how beehives impact the landscape. Most importantly, this data and visualization help advise communities where best to place new beehives as a restoration technique in degraded pastures.

In addition to working directly with villages and community groups, APW partners with local government authorities to facilitate evidence-based conservation decision-making. APW is exploring ArcGIS Hub℠ as a tool to streamline its collaboration with government partners.

In the Monduli District in the Arusha Region of Tanzania, a hub site created using Hub provides access to information on APW programs in 13 villages on topics including human–wildlife conflict, rangeland management, and women's enterprise programs. Rather than navigating through ArcGIS Online or individually sorting various dashboards to see relevant data, a district-level user can view a localized hub that has already been tailored to the community's needs.

The hub provides the district manager with monitoring information on topics such as the number of community members hired for various monitoring efforts, numbers of human–wildlife conflicts responded to, kilograms of honey harvested, and installations of living walls (or green walls) of vegetation in the district.

Metrics are generated using the summary statistics feature in Hub and fed directly from a feature layer created by an active Survey123 form for living wall installations. These dashboards are constantly updated as observations are made, so users can monitor the progress of APW programs in real time.

"This tool can be so powerful for decision-making when put directly into the hands of village- and district-level land managers," said Naro. "For instance, here they can see how many pastures were open for grazing last month, as well as if there were any pastures grazed when they shouldn't have been."

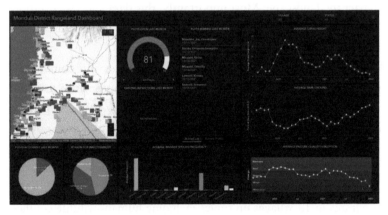

Monduli District Rangeland Dashboard shows plot accessibility and the frequency of invasive species.

APW's efforts also foster local collaboration by supporting neighboring communities in creating equitable grazing plans when critical pastures are inaccessible.

For instance, when Ol Tukai village had four critical plots that were flooded, the rangeland monitoring platform helped the communities and the district government collaboratively visualize data about the affected area. This, in turn, supported the development of intervillage grazing agreements to ensure that neighboring communities share pasture resources for the benefit of all. The APW system provides an agnostic visualization tool to aid in political negotiations by facilitating better-informed and more equitable decision-making for conservation.

Naro illustrated the value of APW's work by sharing the words of a Tanzanian colleague, Yamat Lengai, who works closely with community decision-makers on these types of issues. "In this ever-changing environment where technological tools can be so powerful," Lengai said, "this data will allow us to put our partner communities at the forefront of conservation."

A version of this story by David Gadsden appeared as "Esri Conservation Summit 2021 Blog Series: African People & Wildlife" on the *Esri Industry Blog* on February 15, 2022. The story is based on a user presentation from the Esri Conservation Summit—Africa in September 2021.

ENABLING SELF-SERVICE ACCESS TO VISUALIZE DATA

Open Space Institute

THE VALUE OF DATA AND MAPS MULTIPLIES WHEN THEY ARE readily accessible. When everyone in an organization has secure, straightforward access to data and maps, they are far more likely to use them. This is a logical but not always frequent practice because of system and knowledge limitations.

Open Space Institute (OSI) is an environmental conservation organization and innovative leader in the domain of land trust. By protecting scenic, natural, and historic landscapes, OSI provides public enjoyment, conserves habitat and working lands, and sustains communities. Maps play a key role in land conservation work and are used extensively to inform prioritization planning and communicate progress to stakeholders. OSI had always used the visual power of maps but took a traditional approach by using static paper or

OSI's geospatial hub highlighting its programs.

PDF maps. This approach inundated OSI's GIS team with time-consuming requests for specific maps and data. To respond to its growing needs, OSI implemented ArcGIS Online, a web-based mapping, analysis, and content management system that enabled self-service access to interactive data and maps.

Challenge

A central server in the New York City home office held OSI's geographic information, which staff across the East Coast accessed through a virtual private network (VPN). Remote access was cumbersome and time consuming, which discouraged remote staff from accessing the information. To avoid this process, they would save a copy to their local drive or send an email to the GIS team requesting the specific data and map for their current project.

This was problematic for two reasons: First, the copied geographic information saved to their local drive quickly became outdated and inaccurate. Second, the GIS team spent considerable amounts of time making basic map and data customizations and generating PDF maps.

Solution

OSI implemented ArcGIS Online, which fulfilled the organization's needs through its web-based content management system, powerful dashboards, configurable web apps, straightforward map authoring, and data collection workflows.

Secure, web-based content management system

OSI staff, regardless of where they are located, can conveniently access their data and maps through ArcGIS Online, a software as a service (SaaS) product that runs on any modern web browser. The maps are not static PDF maps; they are interactive web maps that provide enhanced details and new perspectives when zooming in, searching, and interacting with the data.

By organizing their maps, data, and users into groups in ArcGIS Online, the GIS team provides secure and convenient self-service access to OSI's geographic information. Team members, through their licensed identities, are assigned to a group and given access to the subset of the organization's data and maps relevant to their work. The GIS team updates the web maps and data in each group, providing everyone with current information.

OSI staff access their data and maps through a hub site created using Hub, an application included with ArcGIS Online. The site provides an overview and branches into specific pages for each of OSI's programs. The hub site and pages follow the same group sharing rules set up in ArcGIS Online by the GIS team.

Powerful dashboards

To answer commonly asked data and location questions, the GIS team created dashboards using ArcGIS Dashboards, an application included with ArcGIS Online. The dashboards include information about OSI's current landholdings, potential projects, projected dispositions, and grants distributed. Teams including External Affairs and Stewardship use the dashboards to gain a clear snapshot of organizational interests and answer questions such as, "How many easements do we have in Ulster County, New York?" and "How many acres do we own in this state?" The dashboards also provide a quick way to see what OSI is doing in an area of interest.

Straightforward map authoring

OSI staff often have specific visual requirements for the maps they share with stakeholders. These maps may include a specific basemap, symbols of a certain color and size, and a specific map extent. With

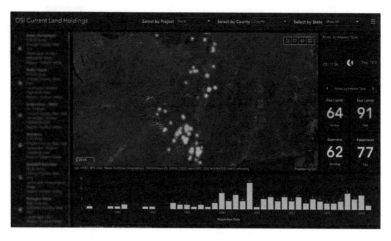

OSI's dashboard showing land types and landholdings in a certain area.

the straightforward visualization tools in ArcGIS Online, users can customize the look and feel of their map. With this workflow, the GIS team can create one map, and the staff can create custom visualizations to meet their specific needs.

Configurable web apps

To share data and collaborate with outside partners, OSI created several configurable applications ranging from simple map viewers to more complex interactive mapping tools using ArcGIS Web AppBuilder. The focus of these applications has grown from singularly focused projects to large, regional landscapes, and they have allowed OSI to work efficiently with partner nongovernmental organizations (NGOs), consulting groups, and federal, state, and local government offices. Depending on the desired experience, OSI configured each application to allow users various capabilities such as filtering and searching datasets, adding their own datasets, and producing simple map exports. All these applications were configured

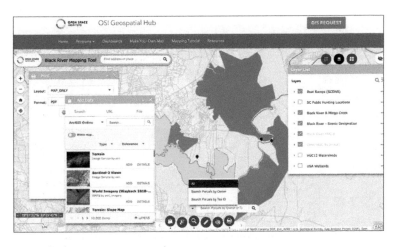

OSI's Black River mapping application.

within the ArcGIS Online system and required no coding, scripting, or special customization.

OSI also built stories with ArcGIS StoryMaps℠ to increase awareness, encourage donations, and gather community input. Its stories include engaging components such as time-lapse maps showing the positive progression of protected lands, guided journeys along river and land, and encouraging narratives of successful conservation efforts. Supporters, local citizens, and partners find OSI's stories through its website, social media platforms, and e-newsletters.

Data collection workflows

OSI brought data from its central server when it implemented ArcGIS Online, but the organization also has an ongoing need to collect data. For example, OSI has a legal requirement to visit every property that it holds an easement on or owns. Historically, after visiting a property, staff would fill out a PDF form and later submit it to the database manager to store. Today, they use the mobile apps included with ArcGIS Online to streamline its workflow. They built property-monitoring forms into Survey123, allowing staff to complete the form, collect the stakeholder's signature, and upload reports directly on-site. ArcGIS Field Maps is used to record site-specific observations, take geolocated images, and track the area and distance covered during each site visit. Data from these applications is fed into ArcGIS Online and can be displayed for instant review.

Results

OSI staff can now efficiently access geographic information. With ArcGIS Online, everyone in the organization, regardless of where they are based, can visualize and answer their geographic questions. They can also share interactive maps on screen in meetings with landowners and partner organizations. This practice clarifies conversations and makes their meetings more efficient.

The OSI GIS team serves an organization of 60 people in multiple offices along the East Coast. Its success illustrates how quickly an organization can transition to web-based mapping, content management, and data collection. ArcGIS Online provides a content and user management system and applications such as Hub and Dashboards that only require configuration—not development—to set up in an organization.

Because the OSI has self-service access to its current data and maps, the GIS team is free to devise more ways to use the technology, and the entire organization is sure to discover new ways to engage its stakeholders through the power of data and maps.

A version of this story titled "Open Space Institute Gives Everyone in their Organization the Power to Visualize Data and Answer Geographical Questions" appeared on esri.com.

USING SMART MAPS TO SAVE LIONS AND SOLVE HUMAN-WILDLIFE CONFLICT

Born Free Foundation

LIONS CLASH WITH LIVESTOCK HERDERS IN SOME REGIONS of Africa, and the situation threatens both. The traditional approach of translocating lions away from cattle often doesn't work. Once lions are moved, many return to prey on livestock, which leads to retaliatory killings by herders. Other translocated lions die from the stress, are attacked by dominant male lions they encounter, or die from other dangers while trying to return home.

Searching for a humane approach to this problem, Born Free works to strengthen cattle corrals, called *bomas*, with tall and sturdy chain-link fencing, making them predator-proof. Traditional bomas made of acacia thorn twigs keep docile cattle in place but cannot keep lions out. For a boma project in the human settlements adjacent to Amboseli National Park in Kenya, Born Free is mapping where fortifications can best help people and wildlife. The maps are built with GIS, which provides location-aware apps for field officers to identify areas that require boma upgrades, based on conflict data, and to measure the impact of past boma fortifications on the conflict between lions and livestock.

As of early 2021, Born Free had fortified 339 bomas, protecting about 100,000 livestock and benefiting almost 7,000 people. The lion population in Amboseli has grown over the years, from as few as 50 individuals in 2008 to about 200 today, thanks in part to the boma program. Better quality of life for nearby herders also signals the program's promise.

"When their bomas were not enclosed, men would spend the night out and weren't able to leave their home for long because they're the main protector," said Linda Kimotho, GIS officer, Born

A lioness near Amboseli National Park in Kenya.

Free Kenya. "But now with a closed boma, you're sure that your livestock is safe. So, it can even free people to travel and go to other areas to look for alternative sources of income to meet their daily needs."

Born Free's conservation managers and scouts help the indigenous population of pastoralists, such as the Maasai, coexist with lions. GIS smart maps help them see where human–predator conflict is reported in the Amboseli ecosystem and where conservation managers should be deployed to help reinforce bomas and peacefully resolve community conflict with lions. In conflict hot spots, GIS helps identify schools and communities where the organization can conduct outreach.

Helping spread the boma solution

Thousands of Maasai people live around Kenya's national parks. During the day, there isn't much threat from lions as herders safeguard their sheep, goats, and cattle; however, nighttime attacks are common.

A lion attack can mean devastating economic loss to a Maasai herder. When community members hear of the attacks, they take measures to stop it—often resulting in death for lions already struggling to survive. Even if the herders respond with peaceful measures, such as staying up all night to watch over the cattle and ward off the lions, the conflict makes it difficult for communities to thrive.

Born Free works with the Maasai to reinforce bomas with chain-link fencing strong enough and high enough to keep lions out. Communities that apply for the boma must pay 25 percent of the cost. Born Free gives the other 75 percent and sends a technical team to construct the boma and train villagers on how to maintain it.

"Communities report any lion sightings to another organization that monitors the lions, communicating where the lions are so that people will avoid taking their livestock there," Kimotho said. "We also encourage herders to embrace better herding practices, such as having more than one herder and engaging adult herders as opposed to children."

The Born Free GIS team has been working on apps for mobile devices that feed a shared dashboard to keep users up to date about progress and incidents in its two main project areas.

"GIS lets us see the distribution of our bomas and know which areas we haven't covered," she said. "We can go there and find out why people in this area are not applying for bomas. Is it that there is no conflict? Is it that they are not aware of the boma program?"

The outreach and location-based research helps Kimotho and the Born Free team learn about lawful practices that sustain local ecosystems. Smart maps give them a way to visualize and analyze the data they collect.

"And if there is no conflict, we are able to also dig more and understand why there is no conflict and why other areas are experiencing conflict," Kimotho said. "Probably it's something the community is doing that can be replicated in other areas."

PART 1: CONSERVATION LAND MANAGEMENT   29

A fence protects a cluster of houses in Kenya.

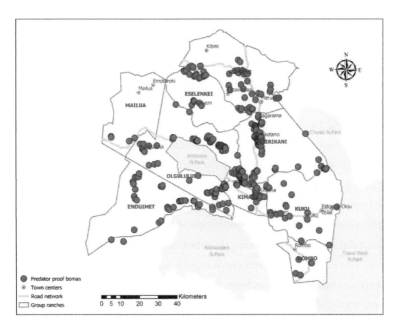

The extent of the Born Free Foundation's work to protect communities throughout Amboseli National Park, including predator-proof bomas, town centers, the road network, and group ranches.

Protecting lions in Meru National Park

The work to help communities live in harmony with wildlife is a new chapter to the "Born Free" ideal launched more than 60 years ago with the publication of the book by the same name. That real-life story by Joy Adamson related how she and her husband, George, saved the orphaned lioness Elsa and raised her so she could return to the wild. The book inspired the 1966 movie, *Born Free*.

Actors Virginia McKenna and her husband, Bill Travers, traveled to Kenya to star in the film and came away with a life-changing experience. Seeing the plight of wild animals inspired them to start the Born Free Foundation for the preservation and humane treatment of wildlife. Today, that foundation continues to fund programs including those in Kenya's Meru National Park, where the original story took place.

"In the home of Elsa, we use GIS to visualize collared and uncollared lions, studying their distribution and home ranges to determine their habitat use and resource selection," Kimotho said. "This helps us understand the current status of lions, identify the major threats that could be causing declines in the current lion population, and develop sustainable solutions to mitigate these threats."

GIS tools have also been used to classify the vegetation in the park, a critical survey that contributes valuable details for the conservation management plan. GIS also contains more habitat variables—such as climatic conditions and topography—to determine such data as animal density, distribution, and habitat use.

The Born Free movement started when large animal populations were decimated by poaching, inhumane zoo captures, development, and wasteful sport hunting.

Today's challenges involve finding a balance between animal habitat and human settlement. That often includes identifying safe, peaceful ways for people to coexist with lions, which are currently listed as a vulnerable species by the International Union for Conservation of Nature (IUCN).

According to IUCN, almost half of Africa's lions have been wiped out in just over 20 years. The lion population in Kenya was 10,000 in the 1980s, and only about 2,000 lions live there today. The Born Free Foundation partners with the Kenya Wildlife Service to help stabilize and grow the lion population in Meru National Park—now home to at least 60 of the big cats.

The foundation also uses GIS to educate the community on sustainable water use and to combat damming of waterways by communities adjacent to Meru National Park. The practice of diversion dams can fragment habitats, cause drought in some areas, and deprive species of water in areas where water has been diverted. The foundation has mapped some of the diversions and is sharing information with local communities on how diversions can impact the health of the river and larger ecosystems, as well as suggesting alternative strategies for using the river water.

"We want to understand the river systems that feed into the park because we realized there's a lot of water diversion," Kimotho said. "We want to understand how they [local communities] utilize their water, and also see the extent of water obstruction, and how we can engage the community to ensure that as much as they also have water needs, they can make sure that the wildlife has enough water in the parks."

A version of this story by David Gadsden titled "Born Free Uses Smart Maps to Save Lions, Solve Human–Wildlife Conflict" appeared on the *Esri Blog* on April 22, 2021.

WIELDING LOCATION INTELLIGENCE TO FIGHT POACHERS

Chengeta Wildlife

CLIMATE CHANGE AND THE NOVEL CORONAVIRUS ARE adding to already difficult wildlife conservation challenges and harsh human conditions across conflict zones in Africa. In response, local ranger teams with Chengeta Wildlife are finding new ways to approach threats to human and wildlife sustainability.

"In places like the Central African Republic, the people literally live hand to mouth," said Rory Young, cofounder of Chengeta Wildlife, a nonprofit dedicated to protecting and promoting harmony between humans and nature. "The work you do feeds you tonight or tomorrow. If they can't work, they are literally starving."

The nexus between poverty, poaching, and wildlife trafficking causes ongoing problems across Africa and around the world. Sadly, Young himself was killed in April 2021 while on a wildlife protection patrol. According to a recent report by the United Nations Office on Drugs and Crime, illicit income generated by elephant ivory is about $400 million annually. Poachers get a small fraction of that take, but in places of extreme poverty the income and the meat provide motivation for opportunistic killing. The activity often takes place outside parks, where there's less risk of detection and a dire need for food.

"I go into dead forests where there isn't a living thing," Young said. "People have eaten the birds, the insects, all the termites. I remember seeing these places thriving with wildlife. There is mass slaughter on the ground, and I don't think that gets conveyed [to a wider audience]."

During the COVID-19 pandemic, networks of criminals have taken advantage of pandemic restrictions that make it difficult for rangers to go on patrol.

Chengeta Wildlife trains rangers on bush tactics, tracking skills, and location intelligence to anticipate, locate, and apprehend poachers.

Organized with information

Chengeta Wildlife maintains a permanent presence in the Central African Republic, Cameroon, the Republic of the Congo, Mali, and Burkina Faso—all countries that have experienced some level of violence.

"We don't specifically seek out conflict zones," Young said. "We go where the need is greatest."

The organization takes an intelligence-driven approach to thwart poaching, training rangers to track and apprehend poachers by anticipating their movements.

"When we deal with a problem in a given area, analytics and actionable intelligence inform our planning, coordination, and execution of missions," Young said. "Improving the technology allows for improved command and control."

Chengeta Wildlife uses GIS-based smart maps and apps to aid rangers and park managers.

"We teach rangers to use GPS and put information on a map to visualize the situation—situational awareness is critical," Young

An example of the proximity of livestock to wildlife (elephants) in the distance.

explained. "The rangers are either collecting data or absorbing the analysis, and they pick it up incredibly fast. They have a keen understanding of geography and appreciate the importance of improving that understanding all the time."

Rangers collect data related to poaching and share it with the Countering Wildlife Trafficking Institute in St. Louis, Missouri. There, geospatial intelligence experts combine all the collected information to target poachers, prioritize missions, and provide tactical support.

"These situations are too complex to be understood in any way other than by visualizing via mapping," Young said. "The better rangers understand their environment, the more they are able to ensure their own safety and succeed in their mission."

Perfect storm in the Sahel

Chengeta Wildlife works in places where humans and wildlife compete for resources, space, and habitat.

Desertification continues to erode arable land in the Sahel region of Africa.

"That's where the real battle's being fought," Young said. "The solution is not just less people or to create more parks. It's really about understanding the conflict generators."

In the Sahel region of Africa, desertification—the breakdown of fertile soil to desert from overuse, climate change, and other human activities—is shrinking the amount of arable land. The Fulani nomadic herders have moved south as the desert has grown, putting their cattle in competition with wildlife and with sedentary ethnic groups that grow crops. Often, these groups have different nationalities, cultures, and religions.

The Sahel is ground zero for climate change because two different environments—rain-free desert and low-rainfall savanna—meet there. The area has seen extreme temperatures, fluctuating rainfall, and decades of drought. The UN Food and Agriculture Organization estimates that 80 percent of the land in the region has been degraded, causing millions of people to suffer from severe food insecurity.

"It should be seen as a prophecy," Young said. "It's going to happen on a larger scale in more places in the world. When the environment is destroyed, people go hungry and they're forced to move, putting pressure on someone else."

Chengeta Wildlife works to understand what's needed. If the problem is hunger, the answer isn't necessarily more arable land. The solution might be to teach new farming practices. If the problem is a lack of water, drilling for water or managing the environment to capture runoff during the rainy season may be the answer.

"Our community experts point out that the problem is often not a lack of money, it's how to manage money and having a place to put it," Young said. "We try to figure out how to help people so that we can help the wildlife."

Progress on the problems

Despite overwhelming challenges, Chengeta Wildlife has had some success. It takes a pragmatic approach—determining where there's need, where people want help, and where the group's solutions will improve the situation.

Rangers receive Chengeta Wildlife's advanced anti-poaching training.

"When we take on a project, we really believe we can make a difference, and then we go for it," Young said.

On one project in the Sahel, Chengeta Wildlife staff have trained hundreds of rangers to protect an area of a quarter-million square kilometers (96,525 square miles). In Mali, staff curtailed the poaching of desert elephants—which had seen a loss of 45 percent of the population from 2015 to 2016—and achieved a zero percent loss by 2019. Now, that region's elephant population is growing, despite escalating security problems.

Young and others at Chengeta Wildlife are trying to repeat a successful pattern. They forge partnerships with like-minded organizations and use GIS to understand the movement of elephants and poachers to predict activity at a particular time and place.

Analyzing all inputs to arrive at geospatial intelligence

Dr. Odean Serrano, founder of the Countering Wildlife Trafficking Institute (CWTI), approaches the problem of wildlife trafficking by applying knowledge from decades of geospatial work for the National Aeronautics and Space Administration (NASA) and the US National Geospatial-Intelligence Agency (NGA). Serrano managed NGA's environmental geography program before joining academia at the Saint Louis University Geospatial Institute.

"At NGA, I analyzed and promoted critical environmental challenges like climate, food, water, biodiversity, and ecosystem health with respect to security," Serrano said. "GIS is the glue that brings together various missions and themes of environmental security to understand who's doing what, when, and where."

Serrano provides the geospatial analytics arm of Chengeta Wildlife's evidence-based intelligence unit alongside an organization called Sensing Clues, which applies investigative analysis.

CWTI and Sensing Clues work together to combine antipoaching

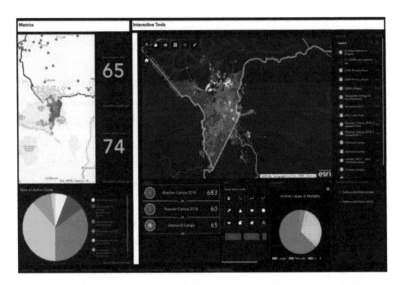

An interactive dashboard displays details about missions, including the locations and total number of poacher camps.

operational analytics with conservation research to provide an understanding of the issues affecting protected wildlife areas. Each conservation operation is informed by a plan with plots on a map of ranger patrols, weather data from satellites, data from GPS collars placed on elephants and gorillas, and specifics about the local population.

Rangers in the field use a data collection and tracking app developed by Sensing Clues, which is seamlessly integrated with ArcGIS Online. These tools allow rangers to record details about what they observe on patrol. The data then gets integrated with historical data to reveal hot spots of illegal activity as well as trends. This information gives the rangers a better sense of what their mission is and where to patrol.

"When you connect an advanced tracker with a skilled geospatial analyst, magic happens," said Young of Chengeta Wildlife. "The ranger knows he can collect information about a group's movements, its composition, and even its weaponry—reading impressions, including every butt of a rifle that's put on the ground. We're linking the ancient skills of trackers with advanced technology and analytics to truly make a difference.

"The geospatial analytics is very important to determine where, when, and how we should move to avoid confrontation or conflict," Young concluded. "It allows us to disrupt the opportunity—getting there, stopping the poaching, and exfiltrating safely to keep our rangers alive."

Chengeta Wildlife honored Young with a tribute after his death.

"From his work as a guide and tracker, in human–wildlife conflict, and then as one of Africa's top strategic trainers, Rory stayed true to his beliefs, applying the at-times tough lessons learnt along the way to make his place in a world that desperately needs passion and dedication," the tribute said in part. "He knew how to bring the best out of people, to inspire them, to remove their fear and replace it with determination."

A version of this story by David Gadsden titled "Chengeta Wildlife Wields Location Intelligence to Fight Poachers" appeared on the *Esri Blog* on November 2, 2020.

HOLLYWOOD TRICKS AND LOCATION TECHNOLOGY CATCH POACHERS

Durrell Institute of Conservation and Ecology at the University of Kent

O N A WARM SATURDAY NIGHT IN 2018, SOMEWHERE IN THE northwest of Costa Rica's Guanacaste province, Helen Pheasey inspected a small white spherical object in the palm of her hand. Pocketing it, she took a short walk to a nearby beach to begin the night's research.

It didn't take her long to find an olive ridley sea turtle, measuring 72 centimeters (about 28 inches) tip to tip, making its slow progress out of the surf and onto the sand. Pheasey watched as the turtle dug an egg chamber with its rear flippers and settled itself above it. Over the next 20 minutes, it laid a clutch of around 100 eggs, the same color and shape as Pheasey's sphere.

When she knew the turtle was halfway through its task, Pheasey

Olive ridley sea turtle.

reached down and deposited the sphere among the eggs. Then she made her way home to wait.

On Monday morning, the fake egg began to move. Embedded with a GPS transmitter, the sphere headed away from the beach deeper into Costa Rica's Central Valley.

On the poachers' path

Every tracking system has three basic components: a GPS receiver to plot location; a communication device, such as a cellular connection, to transmit those location coordinates; and GIS to transfer positions into tracks on a map.

Such systems are usually used to capture the movement of people—by companies or government organizations interested in the efficiency of fleets, for instance—but some are used to track animals.

Pheasey is a conservation biologist at the University of Kent's Durrell Institute of Conservation and Ecology. Her research used a tracking system to monitor two sea turtle species—the green sea turtle and the olive ridley—that have seen a sharp decrease in global population. An estimated 800,000 olive ridley females nest each year, down from a historical estimate of 10 million prior to overexploitation for meat, eggs, and leather. As part of her analysis of Costa Rica's black market in sea turtle eggs, she fooled poachers into picking up the fake eggs.

The decoy, called an "InvestEGGator," was developed by conservation group Paso Pacífico. Kim Williams-Guillén, the group's lead conservation scientist, drew inspiration from an unlikely source. On the TV series *Breaking Bad*, drug enforcement agents attach a tracking device to the underside of a barrel of chemicals used to manufacture methamphetamine, hoping the cargo's path will provide useful information about the drug supply chain.

Williams-Guillén collaborated with Lauren Wilde, a makeup

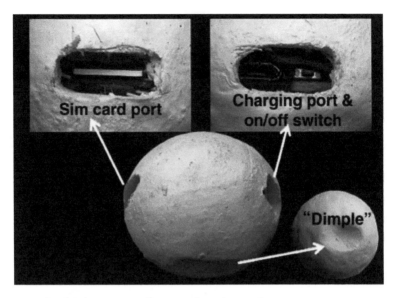

Example of a decoy sea turtle egg with tracker inside.

artist who works with cinematic special effects teams. After studying the composition of sea turtle eggs, Wilde used a polyurethane filament for the housing. A 3D printer creates half an egg. After the electronics are inserted, the second half is printed and fused to the first.

Pheasey learned about the decoy eggs as she was beginning her work in the field in Costa Rica. "The two projects were so compatible," she said. "I'm looking at the illegal wildlife trade, and they're looking for someone to put these things in the ground. They had a prototype ready to scale but no place to deploy it."

Building the perfect egg

Night is when sea turtles lay their eggs, which means it's also the time when poachers ply their trade. A decoy egg can blend into the darkness, which is why tactile sensation is key to its success.

"They look very realistic, especially when they're covered in

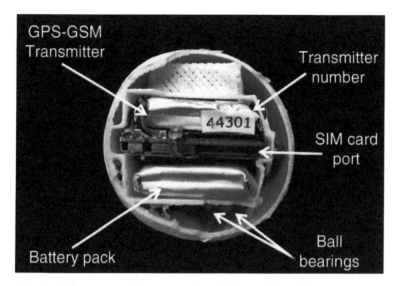

Detailed look inside the tracker placed in the decoy sea turtle egg.

sand at night, but it's even more important that they feel like real eggs," Pheasey said. "A real turtle egg will kind of squish under your thumb."

The eggs even come in different variants, depending on the sea turtle species. For the green sea turtle, the decoys contain ball bearings to make them feel heavier and to approximate the uneven weight distribution of the yolk sloshing around inside.

"They really went to a lot of trouble," Pheasey said. "Fortunately, all I had to do was put them in the turtles' nests."

Will it fool the pros?

Over a period of several weeks, Pheasey and her team deployed 101 decoys. They began on the country's Caribbean coast, where the larger green sea turtle is more common, before moving to the Pacific side to study the olive ridley.

The team's presence didn't draw much suspicion from poachers, who are accustomed to seeing turtle researchers on the beaches at night. Even the deployment of the decoy eggs didn't look suspicious because researchers often reach into the nests to deposit humidity monitors and other sensors.

Nor did the poachers pose much danger to the scientists. Since the payout for this kind of poaching was relatively small—a dozen eggs sold to a trafficker would net about $4—the researchers' work didn't put a huge dent in the income of individual poachers. "It's quick and easy money for them," Pheasey said. The hard-boiled eggs are a popular bar snack. They are also consumed raw in a cocktail called *sangrita* or used in a dish similar to *tortilla de patatas*, or Spanish omelet.

Pheasey knew many of the decoys would not yield useful information. Poachers sometimes spotted them.

Still, the results were promising. Of the 101 fake eggs, 25 were taken by poachers. Five provided trackable signals. Most of the paths ended close to the beach. These were of limited interest. Pheasey was less interested in small-time poachers than in the traffickers who deal in bulk. And one Monday morning, she found what she was looking for.

A glimpse of the supply chain

On a GIS-enabled phone app, Pheasey followed the decoy egg's progress away from the beach. "It just went on and on," she recalled. After traveling 137 kilometers (about 85.13 miles), it stopped moving. Pheasey zoomed in and pinpointed the location as an alley behind a supermarket.

"That back-alley transaction was probably with someone who was then going to sell them door to door," she said. "So you had the whole trade chain and a good indication of the number of players."

PART 1: CONSERVATION LAND MANAGEMENT

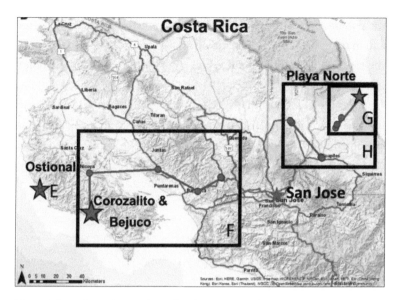

Map of Costa Rica shows where the decoy sea turtle eggs were deployed—and where they ended up.

With the concept successfully proven, Pheasey hoped the decoy eggs would reveal even more details about the larger network. Rather than go after individual poachers, authorities could have a bigger impact by targeting the overall supply chain.

"We want to have enough projects in different locations so we can identify problems and refine the technology, so that it can be a usable tool for law enforcement," she said. "But for now, it's about identifying traffickers and those who are moving large quantities, rather than individuals who are probably from marginalized backgrounds. We're not interested in taking the pittance the poacher is getting. We're more interested in enforcing the law on a larger scale and waiting behind that supermarket loading bay for the next shipment to come in."

More animals could also be in the mix. Pheasey's research has the potential to inspire poaching mitigation efforts for other species and in other locales. Paso Pacífico was even considering how to apply the tracking concept to combat the large black market for hammerhead shark fins.

"The most comparable species to turtles is crocodiles, because in some countries they're poached for their eggs," Pheasey said. But she noted that depositing eggs under a mother crocodile posed some logistical challenges. "It would definitely require a different deployment strategy."

A version of this story by David Gadsden titled "Hollywood Tricks and Location Technology Catch Poachers of Sea Turtle Eggs" appeared on the *Esri Blog* on December 1, 2020.

WHERE THE BATS GO

Bat Conservation International

CAVES HOLD A SPECIAL PLACE IN HUMAN HISTORY. ONCE A common home for people, caves today have become a destination for hikers, spelunkers, and other hobbyists. But they're also used by another set of residents: bats.

Unlike humans, who have mostly abandoned their long-term leases on caves, bats today thrive in certain subterranean environments. And in a twist of fate, society has created a new type of sanctuary for bats: abandoned mines.

Not all old mines make great bat homes, though. Bats require an ideal combination of moisture, temperature, lack of predators, and other factors. Only a choice subset of abandoned mines allows bats to thrive.

That's where Bat Conservation International (BCI) comes in. For

Mobile crews from BCI have visited mines and caves in New Mexico to determine whether they might be bat habitats.

40 years, the organization has worked to prevent the extinction of bats by protecting their habitats. And now, thanks to a new high-accuracy, GIS-based mobile workflow, BCI can locate and safeguard bat-friendly mines more precisely than ever before.

To protect bats, BCI surveys abandoned mines

There are more than 1,400 species of bats worldwide, and although they are critical to the ecosystems in which they live, bats are among the world's most vulnerable wildlife.

In much of the United States, bats are essential to controlling insect populations—their primary prey—and even consume crop pests in large quantities, reducing the need for large-scale pesticide use. In many other parts of the world, bats serve as important pollinators and seed dispersers for numerous fruits, cacti, and other plants—all of which would disappear without the bat species they depend on.

"Bats lead us to the best opportunities to protect nature anywhere in the world," said Mike Daulton, executive director of BCI.

As a collaborative and data-driven nonprofit organization, BCI is largely funded by grants and relies heavily on the aligned goals of federal and regional land management organizations, such as the federal Bureau of Land Management (BLM). BCI's staff of about 30 employees works on preserving 35 critically endangered bat species, 3 of which are found in the continental United States.

As part of an ongoing project that started more than 10 years ago, BCI has worked with BLM to survey thousands of abandoned mines across the western United States to determine which ones offer the best bat habitat. BCI staff members are specifically trained and qualified to safely enter and assess abandoned mines, which present unique hazards and should not be entered by the public. After evaluating the conditions inside a mine, BCI employees learn if and how

PART 1: CONSERVATION LAND MANAGEMENT   49

Abandoned mines now serve as habitats for bats. But not all abandoned mines have the ideal mix of moisture, temperature, and lack of predators for bats to thrive.

bats use the mine so they can recommend specific actions that BLM can take to allow bats to thrive there, such as installing gates that discourage human entry.

"A big part of our work is to survey abandoned mines so that we can recommend appropriate action," said Priyesh Patel, geospatial products and data manager at BCI. "We're contracted to do this for many public land agencies, including the BLM."

In the past, BLM and its regional partners provided BCI with the locations of abandoned mines either as shapefiles or in spreadsheets. BCI would then load this data into ArcGIS Pro to create web maps. Mobile crews could use the web maps to visit and record the conditions at each mine.

But recently, BLM asked BCI to create its own inventory of abandoned mines for a project in New Mexico. Patel, who leads BCI's geospatial work, had to create an entirely new workflow for this.

BCI crew members used Arrow 100 Global Navigation Satellite System (GNSS) receivers from Eos Positioning Systems, along with ArcGIS Collector and Survey123, to validate potential mine entrances.

The process of finding and safeguarding bat habitats

First, Patel used a topographic map of public lands from the United States Geological Survey (USGS) to try to determine known entrances to abandoned mines. After spotting patterns in the maps that indicated probable mine entrances, Patel used ArcGIS Pro to digitize these openings into points. To supplement his topographic map detective work, Patel also referenced satellite imagery, which allowed him to identify more potential mine locations.

The next step was to visit and verify the mines. Patel knew his mobile crew would need submeter accuracy to do this because abandoned mines often have multiple entrances located close together.

"Some mines are clustered together because they follow a common vein of ore," Patel said. "Due to this clustering, it's actually not that uncommon to have two or more mine openings that connect a few feet in. Using cell phone GPS, [which] is often off by several feet

Web maps and dashboards built by BCI show the status of each of the mines that mobile crews inspected.

and sometimes up to 20 meters...wouldn't result in accurate mapping of features, especially when in a cluster. This caused us to realize that we needed to verify each unique mine opening within a few feet."

Patel purchased two Arrow 100 Global Navigation Satellite System (GNSS) receivers from Esri partner Eos Positioning Systems to use for data collection along with Collector and Survey123. The Arrow 100 allowed Patel's crew members to navigate precisely to the points he had digitized and either verify, reject, or update the mine locations. If crew members deemed a point valid, they then used Collector to create the final inventory of the mines and Survey123 to record rich attribute data for each mine, including its temperature, humidity measurements, and whether crew members detected insect parts or guano (bat droppings).

From this data, Patel can easily run a query on the mines BCI crews have visited to determine which ones constitute excellent, good, moderate, and poor bat habitats. Using ArcGIS Online, Patel also created web maps and dashboards that visually display this data so anyone can see the status of each mine.

Mines follow a common vein of ore, so the extent of an underground mine can be vast. BCI verified the area of this underground mine—depicted in red—using lidar data, Arrow 100 GNSS receivers, Collector, and ArcGIS Pro.

Any mine designated an excellent, good, or even moderate habitat for bats is recommended for long-term protection, triggering action by BCI's land management partners. This action often means protecting the mine from human interference, usually by erecting barriers that close off the mine entrances—especially for mine openings that are near roads.

"That simple gesture can lower disturbance and help provide

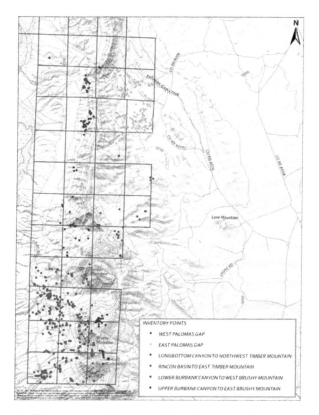

By the end of the project, all the mines BCI had inventoried fit into six distinct geographic regions, represented on this map by different colors.

bats with an undisturbed environment in which they can thrive," said Patel.

Early successes give hope for the future

So far, Patel and his team have verified the locations of 785 mining-related features in New Mexico using their high-accuracy mobile surveys. Of these, about half—or 308 features—have been fully surveyed. Out of those 308 surveyed locations, only 8 showed signs of being live bat habitats. That showed how difficult and critical it is for bats to thrive in unprotected areas.

Initial feedback from BLM has been positive. Patel hopes that in addition to recommending meaningful closures to ideal mines, BCI staff members will be able to revisit the mines years from now to determine if the closures have had a positive impact on bat populations.

"I hope that one day we can [return] to monitor the area and see if this work has helped to improve bat habitat," said Patel.

The team is also using the Arrow 100 GNSS receivers with Field Maps to georeference point clouds, taken with terrestrial lidar, that don't have built-in GPS capabilities.

"It's been fun experimenting with what is possible," said Patel. "Our goal is to get us to the point where land managers can look at data from afar and make good decisions based on the data."

A version of this story appeared as "Where the Bats Go" in the Winter 2022 issue of *ArcNews*.

FINDING BALANCE: HELPING PEOPLE AND PREDATORS COEXIST

Tanzania National Parks

THE UNITED REPUBLIC OF TANZANIA COVERS MORE THAN 945,000 square kilometers (or more than 364,860 square miles). As countries around the world commit to preserving 30 percent of land and water by the year 2030, it's noteworthy that through decades of conservation leadership, Tanzania already manages 30 percent of its land for conservation. This land is organized into several conservation and mixed-use classes. About 10 percent of the land is in national parks, including the iconic Serengeti Plain, a vast ecosystem that is home to zebras, wildebeests, gazelles, and other wildlife species in northwest Tanzania; and Mount Kilimanjaro, a dormant volcano that rises 19,341 feet above sea level near the country's northern border.

National parks in Tanzania support biodiversity preservation and nature-based tourism activities, which contribute significantly to the gross domestic product (GDP) of the nation and to the livelihoods of communities adjacent to protected areas. TANAPA, or Tanzania National Parks, is a parastatal organization established in 1959 to protect and preserve areas designated as national parks for sustainable conservation.

Currently, TANAPA manages 22 national parks covering an area of 104,661 square kilometers across various ecosystems and landforms including savanna, forest, mountains, and wetlands. In September 2021, Senior Conservation Officer Richard Raymond Mbilinyi, head of the TANAPA GIS unit, shared his insights on the use of GIS technology to support sustainable wildlife conservation in Tanzania.

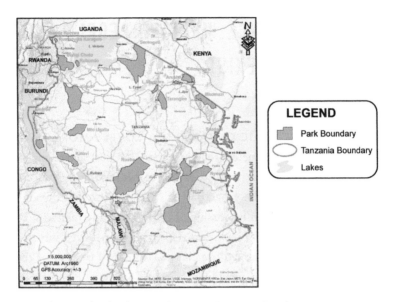

Map showing the distribution of Tanzania's national parks.

The TANAPA GIS unit was established in 2012 with internal sponsorship from the organization's leadership. The aim of the unit is to ensure the day-to-day support of GIS initiatives across the organization to improve operational outcomes. TANAPA currently employs five permanent GIS staff members.

The unit provides GIS solutions to diverse programs across the organization, offering capacity and deployment support in its efforts to sustainably preserve Tanzania's vast biodiversity. TANAPA's 22 national parks are organized into four zones: Northern, Southern, Eastern, and Western. The Western Zone includes more than 10 national parks. Each zone is supported by one GIS officer reporting to Mbilinyi, who leads the unit in the TANAPA headquarters in Arusha, Tanzania.

The GIS infrastructure at TANAPA is powered by ArcGIS Enterprise, with transactions between mobile devices and web apps to

PART 1: CONSERVATION LAND MANAGEMENT   57

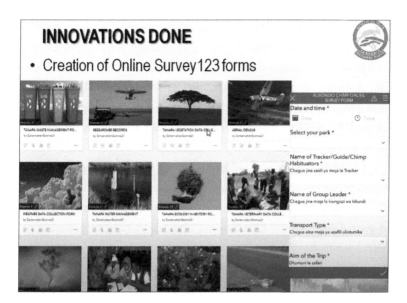

The Rubondo Chimp Online Survey form was created to support wildlife research and monitoring.

support visualization, and decision support provided by ArcGIS Online. The TANAPA ArcGIS Enterprise deployment hosts dedicated geodatabases to support several critical workflows across the national parks, including ecology, protection, tourism, outreach, and administration. As Mbilinyi explained, the following applications provide key capabilities in the TANAPA geospatial infrastructure.

Field data collection

TANAPA collects field data using the modern rugged mobile devices from Trimble running Collector and Survey123. Field data is collected in connected and disconnected environments, which are then synchronized when the device reestablishes a connection to the network. Supporting connected workflows is essential for TANAPA's field teams.

These connected workflows allow the organizations to synchronize their field information in near real time, which has led to several operational advantages, including shorter delays in getting critical information to decision-makers; more secure transactions of sensitive information between the field and managers at the park, zone, and national levels; and reduction of errors and omissions from important field observations.

TANAPA's field data solutions play a significant role in keeping its protected areas safe. Ranger patrols are organized into patrol blocks, which then allows the analysis of coverage and effectiveness of patrol operations. Operational reviews of the performance of patrol teams have led to improved transparency and responsibility through the management chain of command.

For more detailed field surveys, TANAPA has configured Survey123 forms to support ecological monitoring and research on wildlife and vegetation dynamics. Survey123 forms also contribute to the management of physical infrastructure such as buildings and roads, as well as waste management.

TANAPA's field data collection approach and centralized management of mission-critical information in ArcGIS Enterprise have established a single source for authoritative information, which is accessible in focused web apps for decision-makers across the organization.

Web maps and apps

The TANAPA GIS team has developed several innovative web apps that provide data exploration, analysis, and decision support for management teams in the organization. TANAPA web apps feature query, filter, and analysis capabilities, providing managers with a focused data exploration tool to support their work. Web apps are optimized to support each park and each department's workflows, informing their operational decisions daily.

White-bearded wildebeest in TANAPA's Serengeti National Park in Tanzania.

TANAPA makes extensive use of ArcGIS Dashboards, which summarizes patterns and trends from the field into dynamic summary views for managers. When applied to protection operations, TANAPA's dashboards provide quick access to details on poaching incidents coupled with comprehensive patrol metrics from foot and aerial operations to help managers more quickly identify and respond to areas that need protection.

Public engagement

Communicating the vast and awe-inspiring biodiversity in Tanzania's national parks is also supported by the TANAPA GIS unit through ArcGIS StoryMaps. The team has developed storytelling maps for each national park, featuring the parks' scenic beauty, wildlife, and ecology and showcasing the attractions and accommodations available to visitors. A few of the storytelling maps created by the TANAPA team include Serengeti National Park (Western Zone), Ruaha National Park (Southern Zone), Mount Kilimanjaro National Park (Northern Zone), and Saadani National Park (Eastern Zone).

The stories support TANAPA's marketing efforts to attract tourism to the parks while displaying Tanzania's incredible biodiversity as a national heritage and source of pride for the country.

Planning

TANAPA's geospatial infrastructure also provides the basis for long-term planning and management. Many of Tanzania's national parks host significant numbers of the great ecosystem engineer: the elephant. Elephants modify their habitats by pulling down trees and clearing brush, which in turn provides opportunities for other key species who benefit from these disruptions.

Tracking elephant herds and their changes to the landscape informs ecology and protected management operations. It also helps identify potential conflicts with surrounding communities when the herds migrate to other conservation areas. TANAPA has amassed a significant data repository over the past decade, which supports long-term considerations of trends and patterns in ecology, as well as the expanding human footprint outside the protected areas.

As Mbilinyi explained during the Esri Conservation Summit–Africa, GIS technology is crucial for supporting many operational challenges facing TANAPA. TANAPA's geospatial infrastructure provides focused solutions in areas such as wildlife protection, ecology, outreach, tourism, and administration, empowering effective decision-making across the organization.

A version of this story by David Gadsden appeared as "2021 Esri Conservation Summit Blog Series: Tanzania National Parks (TANAPA)" on the *Esri Industry Blog* on January 25, 2022. The story is based on a user presentation from the Esri Conservation Summit—Africa, September 2021.

PART 2

LANDSCAPE CONSERVATION

A GLOBAL CONSERVATION AGENDA CALLED GLOBAL DEAL for Nature aims to conserve 30 percent of Earth's land and water by 2030. The effort, also known as 30 by 30, coincides with Half-Earth, a concept by the late biologist E. O. Wilson calling for us to spare half the planet for nature by 2050. Environmental organizations and indigenous communities around the world are using conservation planning maps and apps powered by geospatial infrastructure as they strive to halt the loss of natural lands and millions of species.

GIS provides the geospatial infrastructure for these efforts by combining advanced analytics, design, visualization, and many other methodologies. This integration provides spatial reference to holistic conservation planning and incorporates analysis and community feedback for sustained landscape conservation. The technology integrates diverse spatial reference data through apps designed for stakeholder outreach and engagement.

Landscape conservation solutions

The discipline of conservation planning has evolved over several decades to identify where conservation actions will have the most impact. GIS today offers an ever-increasing collection of content and tools to advance landscape-level conservation initiatives such as 30 by 30 and Half Earth:

- **Geospatial reference:** Professionals from leading institutions worldwide publish their spatial data with GIS. In so doing, they contribute to the largest and most comprehensive spatial data reference system ever assembled. This vast catalog of best available and authoritative maps, apps, and data layers provides a global canvas for modeling landscape conservation scenarios.

- **Open science:** Analysis and landscape conservation modeling tools native to ArcGIS include on-the-fly image processing and raster analytics. These tools enable seamless spatial analysis with leading open science frameworks such as Python, R, and Jupyter Notebook.

- **Conservation planning:** GIS provides a comprehensive framework for geodesign, the discipline of advising conservation actions through iterative design informed by suitability modeling and endorsed through stakeholder engagement. Geodesign allows users to visualize and quantitatively compare alternative design scenarios for the best outcomes.

- **Outreach and communications:** GIS provides a collaborative framework for exploring conservation scenarios with various stakeholders. Visualizations and storytelling maps inform without overwhelming, and simple survey tools streamline stakeholder feedback.

MAPPING NATIVE FORESTS IN WESTERN CANADA

Conservation North

RESEARCHERS BEGAN TO UPEND MUCH OF THE ACCEPTED science on forests in the 1990s when they discovered that subterranean networks of fungi help trees communicate and cooperate. The symbiotic relationship was first observed when fungi were found to be helping trees transport water and nutrients. Then researchers saw that the fungal threads connecting all trees carry alarm signals and even hormones from tree to tree. With greater awareness of forest interconnectedness—and a realization that old-growth forests are healthier and more tolerant of climatic stress—many industry leaders are rethinking forest practices.

A growing consensus finds that clear-cutting a primary forest—one made naturally and existing for ages—makes little sense. Planting only a monoculture species of a tree isn't prudent either, because growth slows when plantations replace old forests. As tree planting continues on a massive scale around the world, scientists increasingly believe that natural forests can better support our planet through carbon sequestration, the storage of carbon dioxide in forests (a process that can also occur naturally in grasslands, soils, and the ocean).

In British Columbia (BC)—where the awareness of fungal forest communication first took hold—volunteers with Conservation North work to protect wild plants, animals, and their habitats in the central and northern parts of the province. They shared an interactive map of primary forests, intending to preserve them or at least drive awareness of their critical status.

The *Seeing Red* map, created by importing logging data into GIS, revealed that few primary forests remain. Many areas have been

Showing not only the tree canopy but the connectivity of the forest root system.

disturbed by industrial logging, noted in red on the map, whereas shades of green show the untouched forest.

"What shows up as red are the human-created landscapes—industrially managed areas," said Michelle Connolly, director of Conservation North. "The government and industry like to say how sustainable forest management is in BC, and meanwhile, the evidence points otherwise, including the fact that the iconic mountain caribou are basically in an extinction vortex. We're grateful for this tool because it really communicates something completely different from what we've been hearing."

Getting beyond all-green images

To build its interactive map, the team at Conservation North gathered GIS layers from the BC Data Catalogue, the BC Oil and Gas Commission Resource Centre portal, and Canada's National Forest Information System. A report on old-growth forests prepared by scientists served as an impetus.

"The report was a pioneering effort to look at old growth from a provincial perspective that only included tiny static maps, and we wanted to have something we could zoom into," Connolly said. "Both our *Seeing Red* map and *Old Growth* map use the report's methodology, and the scientists who wrote the report have reviewed and approve of our maps—we're proud that they told us the maps are solid and defensible."

The Conservation North team knew it was important to be accurate in their scientific inquiry and mapping because they knew the data revelation would stir emotions. The maps contained indisputable visual evidence many people were unprepared to accept.

Much of the historical logging that took place in the Pacific Northwest was fairly low impact until clear-cutting became the standard practice. The maps detail the more recent story.

"Most of what you see on our maps is the result of modern industrial activity over the last 60 or 70 years," Connolly said. "Our maps are actually an overestimation of what's natural because they don't show all of the historical logging."

While showing the current state of forests in the region, Connolly and her team also wanted the project to help quantify the value of the primary forest and identify where old growth can be saved.

"The exercise started because we needed a conservation planning tool," Connolly said. "We spend a lot of time in the field, witnessing the impacts of industrial forestry. We wanted to know where the largest intact areas are. We could figure that out partly by hiking into these areas, but we needed the perspective only a map could provide."

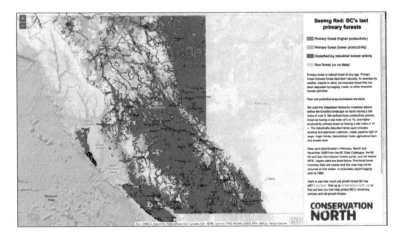

The *Seeing Red* map communicates how much primary forest remains in BC. The definition of primary forest is broader than old growth, because disturbances are natural, and what's left after a fire or insect outbreak is also considered primary forest.

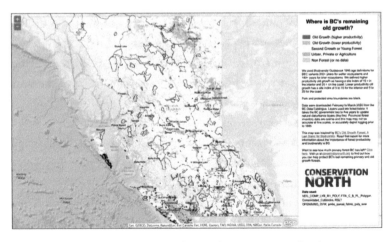

The *Old Growth* map includes only forests that meet an age level. It's a subset map of *Seeing Red* showing the places Conservation North and other activists have tried to conserve for the past 30 years.

The *Seeing Red* and *Old Growth* maps are important to the region's conservation efforts. In 2020, the BC government supported the use of wood pellets—a biomass fuel burned for heating or electricity—but didn't specify that the fuels should come from second-growth forests.

"All forests may now be on the chopping block, and this is terrifying," Connolly said.

Avoiding industrial harm

Conservation North calls itself biocentric—valuing nature for its own sake—with the conviction that people reap therapeutic benefits by just being in a natural setting.

"We recognize that people derive a lot of really important things from forests," Connolly said. "But products should not come from primary forests. They should come from previously logged areas. Sometimes people call this second growth."

In the group's primary focus area in northern BC, Conservation North found three valleys still intact and several more pockets across the landscape they feel are important to protect.

"We know that the biggest, oldest trees are at the most risk from logging, and we've heard that the forest companies have been using lidar to identify all the remaining old-growth forest in the Robson Valley" in British Columbia, Connolly said. "The motivation for our map is to change policy and protect these places for the long term."

Conservation North is focusing on the region known as *Ltha Koh* in the local language of the Dakelh people, which translates to Big Mouth River. This apt description covers an area that sits at the headwaters of the Fraser River, which at about 850 miles long is one of the longest river systems in Canada.

"The Fraser River is a wild river; it's never been dammed," Connolly said. "Ltha Koh is a critical opportunity for protection."

Looking for biodiversity returns

Conservation North volunteers hope their maps can change people's mindsets to foster broader conservation efforts and inspire industry or government leaders to reconsider approaches to forestry.

"There's a pervasive belief within professional forestry that nature needs our help and that humans need to intervene to manage forests," Connolly said. "The map challenges that belief system by showing unmanaged forests as having value on their own. Some places ought to be left alone."

Much of the local industrial pressure in old-growth forests calls for large spruce trees to be harvested as lumber. While the team at Conservation North works to salvage the region's biodiversity, they worry about losing plants and animals that can't be replaced if ecosystems continue to degrade.

"We don't have the towering coastal trees like Vancouver Island and California. We do have a natural rain forest in the interior, and our group was focused on that ecosystem, but we've realized that everything is under threat," Connolly said. "We know that wildlife populations are collapsing across Canada. I've spoken with wildlife biologists who tell me that every species we're bothering to measure is in decline."

Energizing further forest activism

Public awareness of the biodiversity problem has been growing. And when Conservation North's *Seeing Red* map was released, it drew a lot of attention.

"We knew at an intuitive level that industrialization has had a massive impact on the land in BC, but we didn't have a way to see it or communicate it," Connolly said. "Until we did this, no one had a bird's-eye view of just what is left of natural forest in BC."

People and groups from across the province reached out to express their appreciation for the map.

"We heard from community groups and First Nations communities who said the map had a real impact on their work because now they realize what they need to protect," Connolly said. "We're all coming from the perspective that what's already been harmed should be where we harvest, keeping our footprint to those areas and leaving natural areas alone."

Suzanne Simard is the scientist at the University of British Columbia whose PhD thesis vaulted into the prestigious journal *Nature* in 1997. Introducing Simard's groundbreaking research, *Nature* noted that it "shows unequivocally that considerable amounts of carbon—the energy currency of all ecosystems—can flow [...] from tree to tree, indeed, from species to species, in a temperate forest." Connolly took Simard's forest ecology course as an undergraduate, and like most who learn of the language of trees, it changed the way she looks at the forest.

Simard's work to understand forests continues, with a growing focus on resilience and adaptation to climate change. To advance this work for the health of forests and the planet, primary forests must first be preserved. That's where the mapping work of Conservation North could be pivotal—showing where to harvest, protect, and renew—and promote greater carbon capture and biodiversity for flora and fauna.

A version of this story by David Gadsden titled "If Our Forests Could Talk: New Maps Spotlight Forestry Concerns in Canada" appeared on the *Esri Blog* on April 22, 2021.

SEEING MULE DEER DECLINE FROM ABOVE AND CONNECTING THE DOTS

Nevada Department of Wildlife

UNLIKE NEVADA'S LAWMAKERS WHO TEMPORARILY descend on Carson City each legislative session before leaving again, a group of mule deer has taken root in the state capital. It's an increasingly rare sight to see a year-round herd of more than 200 mule deer anywhere in the western United States.

A particularly harsh winter nearly two decades ago severely affected the mule deer population in the state. Its consequences are still being felt today, as 30 percent of the population didn't survive, said Cody Schroeder, a big game biologist and mule deer specialist at the Nevada Department of Wildlife (NDOW).

Ongoing drought conditions in already arid Nevada haven't helped, drying out some of the high-quality vegetation that mule deer need. Then there are other obstacles: grazing competition,

Urban encroachment is one of the threats to mule deer in Nevada.

invasive species, urban encroachment, changing climate, and healthier predators.

To determine the status of the species and to allocate the right number of hunting tags to manage herd sizes, NDOW uses helicopters to tally numbers, gender, age, and health of the mule deer below. Until recently, the surveys often involved logging the data on paper during the flights and waiting until much later to analyze the data. The adoption of GIS changed that.

"Now, we can see right away where deer are concentrated and where we have conflict areas," Schroeder said.

NDOW gathers population details and conducts analysis using GIS technology that allows the agency's biologists to piece together a landscape-level understanding of wildlife and ecosystem health. The data feeds decisions to address the ongoing mule deer decline.

Confronting changing conditions

With recent advancements in how NDOW conducts aerial surveys, decisions on balancing the mule deer's precarious position are informed by a real-time understanding of populations and conditions.

Cody McKee, a biometrician and elk and moose specialist at NDOW, has been working on streamlining data collection, management, and analysis for several years. He was tasked with gathering historical aerial survey data from spreadsheets and filing cabinets and realized that NDOW needed to modernize its workflow to make a leap forward in efficiency.

"Helicopters are an important part of what we do, for that bird's-eye view of the landscape that gives our biologists a holistic perspective," he said. "It's also a dangerous part of our job, and at least for me, the question 'Is this the last time I get into this ship?' is always in the back of my mind. We need to be sure that we are making the most of our time in the air."

Cheatgrass is a winter annual grass native to Europe and southwestern Asia that has become invasive across the western United States.

McKee turned to ArcGIS Survey123 and ArcGIS QuickCapture to create one app on one device for the aerial surveying task and reduce time spent looking down at devices or paperwork instead of looking forward, where hazards lie. The group worked through iterations to improve what had been a paper-based process, using buttons rather than entry fields to standardize observations.

Previously, biologists would take notes and jot GPS points while in the helicopter, and then back in the office, they would spend time typing data into spreadsheets, merging data, and fixing transcription errors. NDOW estimates that biologists were spending half as many hours to get the data usable, so if they spent 1,500 hours flying, it would take an added 750 hours before the data could be analyzed.

Using the app, biologists can tag notes and photos with position and data entered into a shared database. The time saved on manipulating the data gives biologists a chance to reflect. And they can study the data to see trends.

For mule deer, and other species, the data supports queries about the cause of decline.

Feral donkeys increasingly compete with mule deer during drought conditions.

"We used GIS to map the overlap between where mule deer and feral horses are and their preferred habitat," Schroeder said. "We're also looking at other things that are impacting them, such as invasive grass, the drought, and where mountain lions cluster and have kills."

The cheatgrass problem

One of the most vexing problems that Nevada land managers face is the growing impact of cheatgrass—a type of grass that steals water from native vegetation—that has sprouted in large areas where big fires have burned sagebrush habitat.

"Southern Oregon, southern Idaho, and Utah have had a big problem with cheatgrass, but Nevada has had a cheatgrass invasion," Schroeder said.

In winter, mule deer rely on brush sticking up out of the snow. When cheatgrass is present, native shrubs aren't, so deer have nothing to eat.

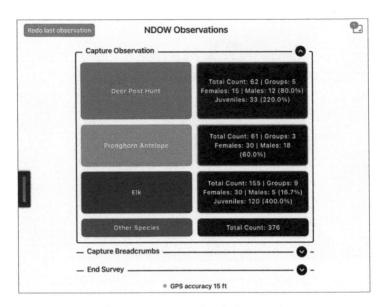

A simple big-button form eases input in the challenging data-entry environment of a helicopter in flight.

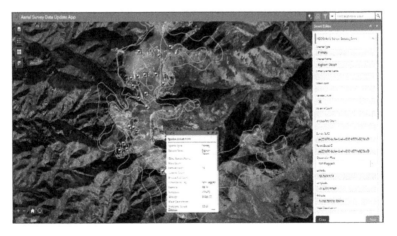

The data of each survey pass (depicted by the blue lines) can be easily displayed on a map alongside the path the helicopter took and what was observed.

PART 2: LANDSCAPE CONSERVATION

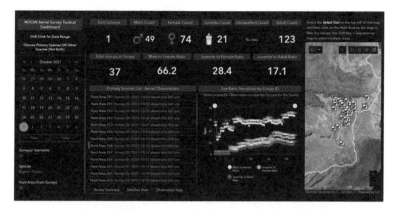

A dashboard view shows details about the status of survey efforts as well as the running total of wildlife counted.

The grass flourishes in Nevada's lower arid elevations, where it has altered whole ecosystems.

"We've documented how cheatgrass changes the natural fire frequencies from 100- to 300-year cycles down to 3- to 5-year cycles," Schroeder said. "It used to be that fires in our ranges were rare, but now they burn over and over."

Feral horses multiply, squeezing out other species

More than 80 percent of the land in Nevada is public, and much of it is managed by the Bureau of Land Management (BLM), which is responsible for herd management of burros and feral horses. BLM sets the Appropriate Management Level of herds and works to keep populations down, noting that an appropriate number in Nevada would be 12,811. However, it estimated there were 42,994 horses and 4,087 burros in the state as of March 2021.

Schroeder teamed with David Stoner from Utah State University and other researchers on a paper about the impact of feral horses on other big-game species.

Feral horses inhabit the same environments as mule deer and compete for forage.

The research involved spatial analysis of the range of varied species such as mule deer, bighorn sheep, elk, and pronghorn antelope, overlapped with the range of feral horses. The paper noted that expanding populations of feral horses are a concern for all species.

Researchers are also using GIS to investigate the impact feral horses are having on waterholes during the drought, on land with little to no vegetation, and what species are displaced and where.

"The research aims to determine how many horses is too many, and where exactly the conflicts arise around water and forage," Schroeder said.

Knowing where to apply more management practices

Biologists use aerial surveys to make many wildlife management decisions because of the perspective they gain from these surveys.

"Once you get up into a helicopter, you realize just how connected everything is," McKee said. "While there are many miles separating mountain ranges, the animals we're managing have the ability to cover those miles in a few hours if need be."

When cheatgrass outcompetes sagebrush, it leaves little forage for mule deer during winter months.

With streamlined data collection, NDOW biologists can analyze pressures spatially and ask geographic questions from the survey data about population health versus range conditions.

"This is going to help us investigate things and focus our habitat restoration efforts where we can create the most connectivity for wildlife," Schroeder said.

Addressing regional mule deer decline

In the mid-1990s, the Western Association of Fish and Wildlife Agencies developed a Mule Deer Working Group that monitors the population across its full range, addresses disease concerns, and fosters best management practices.

NDOW biologists have shared their aerial survey approach with the working group and received great interest. Peers in all states have the same focus on ensuring longevity of species and making decisions that can sustain populations. With all the pressures mule deer and other species face, this group wants forecasts.

"We go up and see this expansive drought-stricken rangeland and know that unless we get the needed precipitation over this winter

Large herds of feral horses disrupt the ecological balance within fragile desert and sagebrush environments.

and spring, our wildlife is going to be faced with some big challenges in the coming year," McKee said.

Future study of Nevada habitat is planned to guide work in places where conflicts cause the most harm.

Biologists place hope in data-driven collaborations to predict and anticipate further catastrophic change. With the new streamlined workflows providing the ability to correlate changing conditions and mule deer reaction, NDOW hopes to engage with other states and stakeholders, including university researchers, to pinpoint causes of decline.

"We don't even know what changes we're going to be looking at in a couple of years," Schroeder said. "But now we can ask these landscape-scale questions."

A version of this story by Mike Bialousz titled "Nevada Sees Mule Deer Decline from Above, Connects Dots with GIS" appeared on the *Esri Blog* on January 25, 2022.

MAPPING AMERICA'S LAND AND SEA: A TIME FOR PRECISION CONSERVATION

Esri

A BIDEN ADMINISTRATION REPORT, *CONSERVING AND Restoring America the Beautiful,* responds to the campaign from several emerging global movements to stop the rapid loss of species and improve resiliency to climate change by setting aside more of the Earth's surface for nature before it's too late. The report proposes preserving 30 percent of US land and oceans by 2030 (30 by 30) through "a 10-year, locally led campaign to conserve and restore the lands and waters upon which we all depend, and that bind us together."

Dawn Wright, chief scientist at Esri, is heartened that the Biden administration recognizes the need for data and science to shape decisions and actions. "As an oceanographer and geographer, I'm also excited because the most critical part of the mission—deciding where to conserve—is one for which we have the technology and approach to support." That specificity can be found in precision conservation, a relatively new methodology that is redefining how landscape and seascape conservation should be approached with smart maps ensuring that conservation projects are of the right size, implemented at the right place, at the right time, and at the right scale.

As the 30 by 30 directive ushers in a new era of needing to work with nature rather than against it, we must reevaluate lands and seas with an eye on biodiversity, carbon emissions, and other environmental factors. We also must be mindful of issues of environmental justice as well as the interests of indigenous populations, local communities, and economic health. To accomplish this, we can borrow from the fields and practices of precision agriculture and precision

Farmers use sensors and maps to determine where interventions can improve crop health and yields.

public health and make advances on precision conservation. These precision approaches are made possible by the range of data collection, management, and analysis supported by a modern GIS, which helps answer questions about the balance between natural and human-made systems.

Knowledge about people, place, and planet enabled by GIS will be crucial to ensure that 30 by 30 actions are environmentally and socially intelligent.

The COVID-19 pandemic showed that science, public policy, and technology can tackle complex challenges to advance the greater good. Interactive maps and spatial analysis can provide greater awareness and context, and sharing of data on a dynamic map allows diverse groups to act together with purpose. As local and national leaders respond to the Biden administration's report, they will need precision conservation to meet the challenges this moment demands.

The location intelligence that can be applied through precision conservation can reveal pathways to smart conservation. For instance, 30 by 30 goals aim to set aside vulnerable and fragile intact ecosystems before they are harmed.

Using GIS, scientists can quantify available places, qualify an area's ability to enhance biodiversity or strengthen climate resilience, and provide critical context to help us understand the role of local and regional ecosystems in the bigger picture.

Meeting the challenge

Just as the agriculture and public health sectors have been using maps and modeling from a GIS, the objectives of 30 by 30 can be guided by GIS technology, which helps us look across space and time to unlock new knowledge about nature and its ecosystems.

Advanced location intelligence from GIS is already empowering a range of conservation initiatives.

Across the globe:

- The Pacific Ocean Accounting Portal spatially integrates public data about the protection, rehabilitation, restoration, and governance of the Pacific Ocean, capturing its real-time condition. It employs the System of National Accounts, which underlies the gross domestic product and other economic measures.

- In the Sahel region of Africa, GIS is used to pick the right species of tree to plant and to monitor the health of each sapling to give the Great Green Wall—an envisioned wall of trees 10 miles wide and 4,350 miles long across northern Africa—a chance of reversing desertification.

- In places such as Palau, Micronesia, detailed dynamic maps show threatened reefs, addressing human activities that put pressure on these vital ecosystems.

In the US:

- The *Map of Biodiversity Importance*, or *MOBI*, was created by NatureServe in collaboration with Esri and The Nature Conservancy. The map provides a comprehensive set of habitat models for more than 2,200 at-risk species, both flora and fauna, in the contiguous United States. It features artificial intelligence (AI) predictor layers that anticipate species viability based on development plans and environmental factors.

- The National Water Model, run by the National Oceanic and Atmospheric Administration (NOAA) and driven by some 7,000 observational measurements, hourly precipitation forecasts, and landscape characteristics, estimates water flow for 2.7 million streams and reaches across the continental United States.

Decades of conservation work around the Chesapeake Bay have started to improve this damaged ecosystem.

- Across the largest US estuary, the Chesapeake Bay, GIS provides a way to communicate and share successful strategies among jurisdictions and states that contribute to water quality.

Today, entirely new kinds of maps and data visualizations are made possible by the instrumentation of natural and human-made systems and the integration of many types of data. These maps and visualizations enable enhanced contextual awareness for better understanding of these complex systems.

GIS is the technology of contextual analysis because it integrates and analyzes diverse datasets and brings the physical world, including people and places, into the core of that analysis to help us see the interconnectedness of our many systems.

Predator fences and traps aim to eliminate invasive pests in New Zealand.

The reefs of Palau, Micronesia, are considered the country's greatest asset.

Changing our mindset

As embraced by the Biden administration, 30 by 30 presents an unprecedented opportunity to unlock and integrate datasets that describe the natural world and enable a decision-support system for preserving biodiversity in the United States.

By applying nature-based conservation strategies, 30 by 30 aims to slow and reverse the environmental degradation that has led to species decline and extinction. This idea of designing with nature has long been a guiding focus of GIS, a technology that works by layering large varieties of data on an interactive map for deeper understanding.

GIS can capture expert knowledge in a single place: a hydrologist's understanding of stream flow, a geologist's understanding of earth structure and process, a botanist's knowledge of plant physiology and classification, an oceanographer's understanding of fish migrations and hurricanes, a policy maker's understanding of implications, and the domain expertise of many more specialists and generalists.

The technology can also integrate with advancements in earth observation and AI, adding more precise and regular measurements and new monitoring capabilities. Drones, for example, provide flexibility, new perspectives, and greater sensing capacity. Anyone can contribute using GIS workflows and a smartphone to capture data and share knowledge about a place.

Sharing maps, models, and GIS-based analysis within the decision-support system of 30 by 30 will help stakeholders make better decisions. Mapping and analysis tools are available to support government leadership as they address this grand challenge.

A version of this story by Dawn Wright titled "Mapping America's Land and Sea: A Time for 'Precision Conservation'" appeared on the *Esri Blog* on May 10, 2021.

MAPS CUT THROUGH THE FOG TO HELP PRESERVE UNIQUE ECOSYSTEMS

Servicio Nacional de Áreas Naturales Protegidas por el Estado

HIGH IN THE FOGGY PERUVIAN HILLS, FOR JUST A FEW weeks in June, bright-yellow flowers resembling tiny phonograph horns light up the hillsides. Known as the Flower of Lima, this plant—with its naturally short, seasonal life and the fog oasis it calls home—is increasingly under threat.

Illegal land-grabbers have seized on an outsized demand for affordable housing as more Peruvians are pushed farther from Lima's established settlements. As a result, land traffickers have sold fraudulent claims on land where the amancae flower (*Ismene amancaes*) grows. The area needs environmental protection and otherwise poses health risks because of its pervasive humidity and vulnerability to earthquakes. The unmanaged development, including unsanctioned

Settlements creep ever closer to the fragile ecosystem.

quarries where miners extract construction materials, has in some cases left behind damaged land unfit for plants.

The unique flora and fauna that make up the fragile ecosystems of this region's *lomas*, or fog oases, have long been at risk of disappearing. In late 2016, the Lomas EbA project, an initiative directed by the National Service of Natural Areas Protected by the State (SERNANP) of Peru and implemented by the United Nations Development Program (UNDP) with support from the Global Environment Fund (GEF), was created to preserve 19 of the estimated 100 lomas that occur only along the desert coasts of Peru and northern Chile.

"We can't lose the only place where this amazing flower grows," said Adriana Kato, a communications specialist with the UNDP group working to preserve the lomas.

Map-based activism

The disappearing fog oases in this high-elevation desert are at the crux of a demand for housing, with about one-third of Peru's

This photo of lomas in Lima's San Bartolo district was taken in August 2019 during the wet season, with the amancae in bloom.

Defenders of the lomas gather to illustrate how this unique ecosystem contributes to the UN Sustainable Development Goals.

population, nearly 11 million people, living in Lima. While poverty levels in Peru dropped from nearly 59 percent to about 22 percent between 2004 and 2018, much of the population remains on the cusp of falling back into poverty. With a lack of housing strategy in Lima, unorganized development has encroached into the fog oases.

"It's not just an environmental problem," Kato said. As a recent UNDP report noted, the effort is aimed at addressing an "unprecedented complex combination of problems."

With so many living in Lima, one might assume it would be easy to find environmental defenders. But the hills faced a unique challenge: few people have witnessed the land in its full glory because it's covered in a blanket of mist from May to October. By the time the fog lifts, the land has turned back to dry desert. Kato, who grew up in Lima, was among those who didn't know about the nearby beauty until she joined the UNDP.

"Lima is surrounded by these fog oases, these green hills, but not many people know about them," Kato said. "We have all this, but we don't see it."

Volunteers use maps to raise public awareness. Photo courtesy of *Haz tu mundo verde*, the volunteer group pictured.

The UNDP's María Miyasiro, a GIS and remote sensing specialist, started raising awareness by building a detailed geoportal to share local apps and maps, including one named GeoLomas to allow anyone to explore the areas that had gone unseen. Local environmental leaders, often self-taught experts in botany and laws relating to the lands, have gathered data for GeoLomas using the Survey123 app on phones and tablets to input what they observe as part of a pilot project.

"I decided to build a web app so our stakeholders and our environmental leaders could do their own analyses and make their own maps," Miyasiro said. "We started with just one application, the GeoLomas web map viewer, and then we elaborated with more applications created together with the stakeholders, organizations, and municipality workers, and using ArcGIS StoryMaps to publicize their work."

Volunteers have been adding layers to the GeoLomas map, including archaeological and cultural features and potential tourism locations, as well as land rights to indicate when an area is being

Noe Neira, a leader of *Lomas de Paraíso*, uses a map to show what the group is defending.

illegally encroached on. The group worked closely with park rangers to record flora and fauna species they encountered.

The online map was described in a recent midterm review of the project as the biggest advance toward tracking conservation actions and improving transparency.

Sustainable development around Lima

The UNDP team also set out to tackle several of the UN's 17 Sustainable Development Goals: improving the environment for those living on the edge of Lima where there has been less access to green space, creating a sustainable and resilient city, conserving the land and biodiversity of hill ecosystems, and reducing poverty by creating jobs in ecotourism and agriculture.

"It's really necessary that people understand that the lomas are not the place to put their houses," Kato said. "It's really dangerous because of earthquake risk and for their health because of high humidity. It's better to conserve them, and make a really nice and

PART 2: LANDSCAPE CONSERVATION   93

Nataly Julca Terrones, a geographer and volunteer, uses the GeoLomas map viewer on her phone to understand the border of the ecosystem.

Municipal workers are trained on how to use the apps on the *Geoportal de las Lomas de Lima*.

Viscachas, native to the area, look similar to rabbits but are not closely related.

safe path, so people will visit them and use it for ecotourism, encouraging community members who may have otherwise lived in the oases to instead preserve the lands and launch enterprises for paying visitors."

In an audit of the lomas project's progress, a report noted that despite the complexities faced, "the balance sheet is positive in terms of will to participate." In Lima, about 10 different local organizations care for nearby fog oases, in some cases starting nurseries to nurture new plantings and using a fog-catching mechanism to collect water for irrigation.

Before the creation of the online map, many didn't know the physical boundaries of the land they were protecting. Now, on a smartphone using the ArcGIS Explorer app or directly in the browser using the GeoLomas web app, they can see whether a house has been illegally erected inside the border or if a new road and plots for sale belong to land-grabbers, requiring a call to the police.

During the pandemic, outreach and training for contributions to the GeoLomas map shifted even further online, and leaders have

The wild potato flower is another common sight during the wet season.

leaned on more technology. Drone footage of the land has become a primary feature on the maps, helping to raise awareness with immersive video that allows anyone to fly over the scenic hills.

In a little over four years, the work of organizers and the volunteer force of residents helped map and catalog crucial information about this area. Those details provided the evidence needed for formal preservation of at least five of the oases, which are now part of a newly designated regional conservation area called *Sistema de Lomas de Lima*.

The UNDP has devised a strategy for handing control of conservation efforts back to Lima's government, including officials from housing, cultural affairs, and law enforcement.

The team encourages private conservation developments and works with the Ministry of Culture to protect other fog oases that include archaeological features at risk of being lost. The team has also partnered with the Geological Institute of Peru to preserve unique geologic formations.

By collecting key data and mapping it with GIS, advocates

Burrowing owls also make their home on the lomas.

communicated the location of each fog oasis, identified local flowers and animal species, noted nearby archaeological locations, and offered details about pedestrian access.

"It's a really, really valuable tool for us," Kato said, recounting what GIS revealed about the lomas. As she and Miyasiro wrote in the interactive map they authored, "You don't take care of what you don't love, and you don't love what you don't know."

A version of this story by David Gadsden titled "Maps Cut through the Fog in Peru to Help Preserve Unique Ecosystems" appeared on the *Esri Blog* on May 18, 2021. All images courtesy of UNDP Peru unless otherwise noted.

WEB MAP BRINGS TOGETHER CONSERVATION AND GREEN ENERGY DEVELOPMENT

The Nature Conservancy

THE MIDWEST IS KNOWN AS THE WIND BELT OF THE UNITED States, and for good reason: nearly 80 percent of the country's current and planned wind energy capacity exists in the Great Plains, an area that extends east of the Rocky Mountains and runs from northern Montana to southern Texas. Wind energy shows tremendous potential as a clean, renewable energy source that can help reduce greenhouse gas emissions.

Much of wind energy development is occurring—and is expected to increase—in the wind belt. But as wind energy developers plan new sites, they face this question: How can new wind turbines be sited in places that are optimal for wind resources and transmission yet aren't likely to impact wildlife or encounter costly delays from regulatory or legal challenges?

Wind projects sited in the wrong place can threaten some of the best wildlife habitat. The Nature Conservancy (TNC) estimates that renewable energy development could adversely affect as much as 76 million acres of land in the United States—an area about the size of Arizona.

But a new GIS-based resource developed by TNC can help focus renewable energy in the right places—windy areas that pose a relatively low risk to wildlife and their habitats. Called *Site Wind Right*, this interactive online map is available for use by wind developers, power purchasers, utilities, companies, state agencies, and municipalities to help reduce conflict between wind energy and conservation.

TNC developed Site Wind Right for 17 states in the Midwest, pulling from more than 100 datasets on wildlife habitat and land use to help highlight areas with the lowest potential for environmental friction. The results of this analysis, done by TNC scientists, are both enlightening and encouraging.

"We were thrilled to discover we could generate more than 1,000 gigawatts of wind power in the central [United States], solely from new projects sited away from important wildlife areas," said Mike Fuhr, state director of TNC in Oklahoma. "That's a lot of potential energy, comparable to total US electric generation from all sources today. While advancements in transmission and storage would be needed to fully realize this wind energy potential, it proves we can have both clean power and the land and wildlife we love."

Great potential for wind in the Great Plains

What eventually became the Site Wind Right analysis started evolving in 2011 for two reasons. First, wind energy facilities began to operate across the Great Plains. Second, TNC and other scientific studies demonstrated considerable potential for wind and solar energy development in the western and central United States.

The Great Plains is home to the country's largest and most intact temperate grasslands, and yet it is one of the world's least protected habitats. In the Greater Flint Hills ecosystem of Kansas and Oklahoma, poorly sited wind turbines have seriously threatened wildlife that depend on this endangered and beautiful place that is home to bison, bald eagles, and the once-common greater prairie chicken.

But as studies demonstrated, the Great Plains could provide clean, renewable electricity that doesn't compromise wildlife habitat and other natural resources.

"Those studies showed very positive results that we can meet or exceed renewable energy goals by using sites that were previously

PART 2: LANDSCAPE CONSERVATION   99

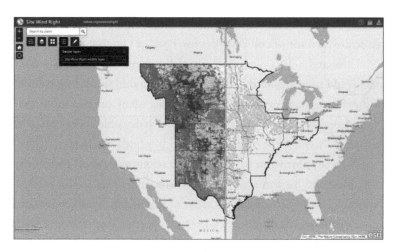

TNC's Site Wind Right, which evaluates more than 100 datasets from 17 states, shows that 90 million acres in the US wind belt could be developed for wind energy without affecting key wildlife habitats, which are depicted on the map by the varying colors. Map courtesy of TNC.

The temperate grasslands of the Great Plains—home to bison and other wildlife—are among the world's most altered and least protected habitats. Photo courtesy of Chris Helzer, TNC.

disturbed or had relatively low conservation value," said Chris Hise, associate director of conservation for TNC in Oklahoma.

Ultimately, TNC scientists created a resource that energy planners could use early in the siting process to avoid impacting wildlife and delaying their projects. TNC is among many organizations that want properly sited wind, solar, and other renewable energy projects to succeed to meet the challenges posed by climate change.

With support from partner organizations and other TNC scientists, Hise and his team collected vast amounts of data—on wildlife, habitats, land-use restrictions, areas of significant biodiversity, and more—and organized it in ArcGIS Desktop using ArcCatalog™. With ArcMap and ModelBuilder™, the TNC team then assembled multiple spatial data layers of wildlife habitats and potential engineering and land-use constraints. Finally, using Web AppBuilder, the team created an online resource that could share this data in what became the Site Wind Right interactive map.

Hise and his team found an impressive number of low-impact areas across the central United States in the analysis—approximately 90 million acres. Planners in the early stages of establishing a wind energy operation can see site-specific details, explore Site Wind Right, consult with appropriate state wildlife agencies, and use the Wind Energy Guidelines developed by the US Fish and Wildlife Service to find spots that work best for everyone. And the low-impact sites in the Midwest are very well distributed.

"If we plan carefully, there's plenty of space to go big on wind energy in this part of the country," said Hise.

Broadening the reach of wildlife-minded green energy projects

Site Wind Right has the potential to reduce the risks of wind deployment delays and cost overruns by helping developers locate sites that

The once-common greater prairie chicken population has fared poorly as its grassland habitats have been converted to other uses. Photo courtesy of Harvey Payne, TNC.

are less likely to face regulatory or legal challenges. This has spurred the endorsement of Evergy, an energy provider in Kansas and Missouri that became an early user of the analysis.

"Site Wind Right is an invaluable resource that helps us avoid unnecessary impacts to the wildlife and iconic landscapes of the Great Plains while also allowing us to provide clean, low-carbon energy for our customers," said former Evergy CEO Terry Bassham.

The mapping analysis invited accolades from another early reviewer, the Association of Fish & Wildlife Agencies, which conferred its 2019 Climate Adaptation Leadership Award for Natural Resources on Site Wind Right. Additionally, the web map has received endorsements from several conservation groups, including the National Wildlife Federation and the Natural Resources Defense Council.

"We need more resources like this to speed up our move away

from burning fossil fuels," said Katie Umekubo, a senior attorney at the National Resources Defense Council. "Well-sited wind energy allows us to meet our climate goals, advances conservation, and ensures that we avoid irreversible environmental impacts."

Currently, TNC is looking to broaden the reach of Site Wind Right within communities, companies, and government agencies so they can quickly apply this wildlife-minded strategy and get the wind turbine blades turning on clean and homegrown energy in the Great Plains.

"The Nature Conservancy supports the rapid acceleration of renewable energy development in the United States to help reduce carbon pollution," said Fuhr of the TNC. "We are looking forward to providing Site Wind Right to the people making important decisions about our nation's clean energy future."

A version of this story by Eric Aldrich titled "Web Map Brings Together Wildlife Conservation and Green Energy Development" appeared in the Fall 2020 issue of ArcNews.

SCIENTISTS COLLABORATE TO MAP BIODIVERSITY AND THE HUMAN FOOTPRINT

Alberta Biodiversity Monitoring Institute

EVERY YEAR, THE ALBERTA BIODIVERSITY MONITORING Institute (ABMI) sends about 60 field technologists across the province to collect samples of biodiversity. These researchers measure habitat characteristics at monitoring locations from a province-wide grid of 1,656 randomly selected sites.

For more than a decade, this unique undertaking has involved ABMI field and laboratory staff entering data into a GIS about the characteristics of habitats and species of plants and animals they find. Many governments have talked about doing an inventory, mapping, and monitoring effort at this scale, but few have done it. Thanks to the long-term commitment of the Government of Alberta and a consortium of scientists, knowledge of biodiversity and human impact continues to grow.

ABMI's appetite for data encompasses everything from soil microbes to apex predators. ABMI equips its field technicians with mobile devices and trains them in scientific sampling methods and backcountry skills. The months-long field deployment may involve daily commutes by helicopter to get to the remote areas where samples are needed.

Since its inception in 2007, ABMI has captured measurements in all the different ecosystems across the province of Alberta, from the grasslands in the south to the boreal forests in the north. ABMI collects data on a range of species (plants, animals, birds, and more) and maintains records on some 2,500 species. More than 1 million

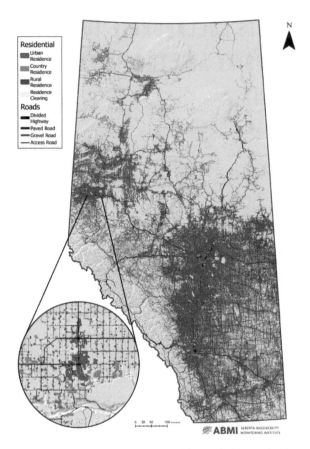

Roads and residential areas are just some of the many human land-use changes that ABMI tracks for the Alberta human footprint inventory.

specimens have been processed, yielding discoveries of several species new to science.

The resultant data is used in a variety of ways to formulate regional and subregional plans, evaluate development impacts, and fine-tune sustainable practices. Local governments use the data to guide urban expansion.

Collecting the facts to mark change

ABMI is a scientific partnership between InnoTech Alberta, the University of Alberta, and the University of Calgary. The biodiversity work had been ongoing for seven years when funding from the Alberta Ministry of Environment and Parks (AEP) established the Alberta Human Footprint Monitoring Program (AHFMP). The program is a joint effort between AEP and ABMI to map and monitor land transformation and assess the impact of human development over time.

"There was a lot of discussion early on about how we brand the initiative," said Jim Herbers, executive director of ABMI. "Words like *human disturbance* would automatically alienate half the people we want to work with. So, we landed on the term *human footprint*. It talks about the activity, the changes to the landscape, in a value-neutral way that is inclusive."

There are sensitivities in Alberta—as anywhere—about capturing the facts and not coloring them with a bias that favors or disfavors any one industry or people, including indigenous groups.

"We need to understand everything that's happening on the land surface, from natural dynamics to land-use-driven changes," said Jahan Kariyeva, director of the Geospatial Centre at ABMI. "Although we're not making or proposing land-use decisions, we use the data to assess how the habitat has changed."

Including the human footprint in ABMI's province-wide monitoring efforts meant adding data in GIS that denotes human activity—reservoirs, roads, railways, mining, timber harvesting, oil and gas, and industrial sites. Prior to AHFMP's creation, the human footprint was only captured for project sites and a few study areas. There was also a lack of sustained monitoring and standardization regarding what constitutes the human footprint.

Helicopters, ATVs, and an amphibious all-terrain vehicle called the Argo are used by the field team to reach monitoring locations. Photo courtesy of the Alberta Biodiversity Monitoring Institute.

Fit for all purposes

From the oil and gas sector, AHFMP has gathered the footprint of more than 350,000 well pads and 308,000 kilometers (about 191,385 miles) of pipeline corridors. Users of ABMI data can examine these features to gain an unbiased picture of how humans have changed the land over time. The data supports all questions and eliminates the more nuanced record keeping of different domains, such as how a forest access road might be recorded. In the AHFMP database, evidence of the road will always remain even if it is now unused and overgrown.

"We work with stakeholders to see if we can incorporate their data," Kariyeva said. "If it's not possible or feasible, we reach out to subject matter experts to verify and validate our work."

ABMI makes the data available for free in multiple formats to be explored and queried by domain experts, policy makers, and anyone interested. For everything it records, ABMI provides the raw data that anyone can search and download by species and habitat.

ABMI's sampling sites are kept private to protect the science. This photo of the historic site of the Head-Smashed-In Buffalo Jump shows the typical southern Albertan prairie and depicts an early human use of the land, where bison were killed by being driven off the cliff.

In addition to data of what's on the land, ABMI has developed several province-wide data products about the land that can be explored using GIS. These products include map layers about soil and climate and a recently completed wetland inventory.

"ABMI and Ducks Unlimited Canada developed a partnership and recently invited Alberta Environment and Parks to join us," Kariyeva said. "Together, we all work toward one single wetland inventory that supports everybody's needs."

Classifying location data to ask questions

ABMI turns its data into knowledge through products such as the *Status of Human Footprint in Alberta* report. The data supports queries about the impacts of different industries on different types and classes of species. It's also used to perform cumulative impact assessments for specific industries.

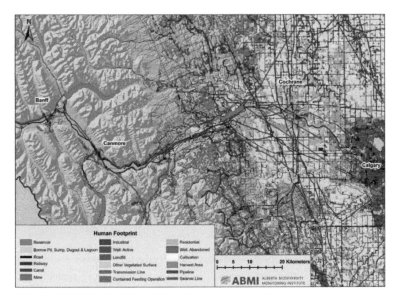

The map of the human footprint in Alberta captures agriculture, forestry, transportation, human-created water bodies, urban and industrial uses, and energy features such as mines and wells.

A recent project assessed the impact of beef production on biodiversity in Alberta. Other projects take longer, such as monitoring caribou and rare animals in Alberta's boreal forest. In the ecological recovery realm, ABMI assesses regeneration of disturbed areas to understand how quickly the forest recovers after disturbance.

The data helps answer many questions about human activity and includes the ability to isolate and remove human activity for greater insight into natural processes and disruptions. ABMI's Biodiversity Intactness Index compares the concentrations of species across regions to predict the abundance of species with no human footprint.

In a recent development, the data may inform several subregional plans that involve broad stakeholders, including industry and indigenous groups. The effort involves gathering more granular data about the human footprint.

Through the ABMI effort, and the partnerships across multiple stakeholders, Albertans have gained the ability to answer questions about biodiversity and how the human footprint has affected their surroundings.

A version of this story by Ryan Perkl titled "Canadian Scientists Collaborate to Map Biodiversity and the Human Footprint" appeared on the *Esri Blog* on August 31, 2021.

CONSERVING A NETWORK OF CLIMATE-RESILIENT LANDS

The Nature Conservancy

THE RESILIENT AND CONNECTED NETWORK, A GIS MAPPING tool developed by more than 270 scientists led by The Nature Conservancy (TNC), gives conservationists a way to save biodiversity for the future. The tool is available for land trusts, government agencies, and scientists for conservation planning throughout the contiguous 48 US states.

GIS has long been one of conservationists' main tools to identify and protect land, water, plants, fish, wildlife, and habitat. Using GIS to identify places with high levels of biodiversity has often played a leading role in those efforts.

Western Iowa's Loess Hills harbor diverse plants and animals. This area holds more than half of Iowa's remaining tallgrass prairie habitat. Photo courtesy of Chris Helzer, TNC.

Climate change is altering biodiversity and habitats. "As warmer temperatures, increased flooding, and other climate impacts alter and destroy habitat, species are being forced to find new places to live," according to TNC. "Diverse plants and animals that thrive in one landscape today may end up living in a very different landscape in the future."

Scientists have found that in North America, species are moving an average of 11 miles north and 36 feet higher in elevation each decade to find more hospitable places to live.

Natural neighborhoods and highways

Mark Anderson, director of science for TNC's Eastern US Division, has studied the connections between climate change and shifting biodiversity for many years. Anderson is also a GIS user who understands the crucial role this technology plays for land trusts, state agencies, and other entities planning conservation projects. Conservationists have limited funding and face huge challenges in their work preserving biodiversity and clean water and protecting landscape health. They need GIS to help identify and focus on high-priority projects.

Anderson wanted to see how GIS could identify places that would have long-term resilience as the climate changes. The TNC reported that Anderson's team found that steep slopes, tall mountains, and other landscapes with diverse physical characteristics can create microclimates in which plants and animals can find habitats to escape increased temperatures, floods, and drought.

Anderson and his colleague Melissa Clark mapped natural highways across the country—connecting corridors that allow species to move safely within and between these climate-resilient neighborhoods.

Although studies show that species are moving to cooler places, nearly 60 percent of US lands and waters are fragmented by human development, blocking the movement of species and preventing them from finding new homes.

"Nature is on the move," Anderson said. "It's not enough to have isolated and disconnected landscapes that are resilient to climate impacts. Species also need a way to reach these resilient sites. While some species will be able to relocate to new homes within their local resilient neighborhoods, others will need to move great distances to entirely new landscapes. If these pathways are destroyed, many species could disappear forever."

A ton of data and essential tools

To create the Resilient and Connected Network mapping tool, Anderson and his team had to "compress 200 gigabytes of data into a useful tool and make it look and work seamlessly," said Erik Martin, a TNC spatial ecologist who helped lead the GIS work. "The map reflects a ton of data, pulling from hundreds of datasets from all around the country."

Using advanced spatial analysis and modeling tools, TNC scientists Clark, Arlene Olivero Sheldon, and Analie Barnett developed, tested, and retested methods to map and measure characteristics of resilient lands. Their analytical and problem-solving skills, coupled with their determination and passion for conservation, took the idea of a network of resilient lands from concept to reality.

One of the big challenges, according to Anderson, was persisting during the 10-year project life span. During that time, the project grew region by region, starting with the Northeast. To ensure accuracy as the map was being built, the team had 270 scientists look at localized geography or geography at a nationwide scale, depending on their expertise. Aiding in that process was the ability to share regional versions via ArcGIS StoryMaps.

Many different ArcGIS products and related apps supported their work. ArcGIS Server, for example, hosted dynamic map services and geoprocessing services that ran custom scripts based on ArcPy™. Many components came together, making the map fast and responsive for users.

The team relied heavily on ArcMap and ArcGIS Pro, using 30-meter scale raster data of the entire country for analysis and mapping. ArcGIS Spatial Analyst™ contained useful toolsets. ModelBuilder was used to help test analysis and iterate steps in each region. Getting feedback from regional scientists helped shape the final product. And ArcGIS StoryMaps stories were key in explaining analysis results and highlighting resilient places across the United States.

"Each of these components is like a building block of a house. We used the ArcGIS API for JavaScript™ to build the frame of the house—the structure of the app. Then [we] connected together all

Over the past 10 years, TNC scientists mapped a network of landscapes across the United States with unique topographies, geologies, and other characteristics that can help species withstand climate impacts.

of the electrical and plumbing—the map and geoprocessing services that feed into the app," Martin said. ArcGIS tools integrated the data into one useful app.

A paradigm shift for large landscape conservation

When the nationwide map was completed, one of the big surprises for Anderson was the large number of areas around the country that are resilient to climate change. "When I first envisioned this, I thought there would be a bunch of big, broad areas. It's really like a web, or a network of connected places."

The map revealed many large and small resilient connected lands:

- Nevada's Monsoon Passage, a natural highway of mountain ranges and wet valley bottoms that extends from Lake Mead through the Great Basin National Park to the Idaho border.

- The Cumberland Forest, which spans 253,000 acres across Tennessee, Kentucky, and Virginia, safeguarding wildlife habitat and storing millions of tons of carbon.

- Wisconsin's Kettle Moraine region, which was shaped by glaciers and contains diverse features such as large kettle lakes and 300-foot-high ridges.

- Bobcat Alley, a 32,000-acre forested corridor in northwestern New Jersey that provides habitat to state-endangered bobcats.

Along with providing safe places where species can thrive, the network of resilient lands also benefits people. The lands mapped in the eastern United States, for example, contain 75 percent of the region's sources of drinking water, generate billions of dollars in outdoor recreation, store an estimated 3.9 billion tons of carbon, and mitigate 1.3 million tons of pollution. Taken together, this results in an estimated $913 million in avoided health-care costs.

In the area near Oklahoma's Canadian River, prairies and contrasting wooded canyons provide climate-resilient habitat for plants and wildlife. Photo courtesy of Jay Pruett, TNC.

"To achieve conservation at the scale needed, we must collaborate with people and organizations across both the public and private sectors," Anderson said. "Keeping these resilient areas safe and healthy will require a wide range of conservation practices including such things as sustainable management, public and private land acquisition, and easements."

Numerous state and federal agencies and land trusts have already incorporated the map's data into their conservation planning. Among the growing number of states incorporating the map into their state wildlife action plans are Connecticut, Massachusetts, New Hampshire, New York, and South Carolina.

The urgency of protecting lands and biodiversity given the challenges of climate change is clear and pressing, according to Andrew Bowman, president of the Land Trust Alliance. The Resilient and Connected Network is a useful tool for that task.

"This map provides a new way to look at the landscape for purposes of prioritizing lands for the conservation of biodiversity,"

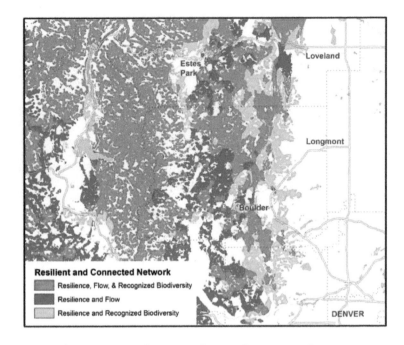

Diverse characteristics such as steep slopes, tall mountains, deep ravines, and diverse soil types help make places resilient to climate change. This diversity creates numerous microclimates and allow plants and animals to move to suitable habitat. The landscapes also have good connectivity, allowing movement across local microclimates and between climate-resilient landscapes.

Bowman said. "I've watched this work take shape over the last decade, and I truly believe it marks a paradigm shift in how we think about large landscape conservation."

A version of this story by Eric Aldrich titled "Conserving a Network of Climate-Resilient Lands" appeared in the Winter 2021 issue of *ArcUser*.

NEXT STEPS

The Geographic Approach to conservation

THE GEOGRAPHIC APPROACH TO CONSERVATION IMPROVES our understanding of the complex web of threats, opportunities, and challenges facing our natural world. In that context, Web GIS provides a way for conservation management systems to unlock research data and support conservation managers as they engage surrounding communities and adapt to dynamic circumstances. This section offers recommended steps to help conservation organizations get started with GIS.

Links for these and additional resources to support your journey can be found on the web page for this book, go.esri.com/pop-resources.

Explore ArcGIS Living Atlas of the World

A vast collection of over 10,000 authoritative maps, apps, and reference data is available in a ready-to-use, curated content from ArcGIS Living Atlas of the World. Explore ArcGIS Living Atlas for topics of interest to learn more about your focus areas and build your own collection of favorite content to support your work. ArcGIS Living Atlas is constantly updated with remote sensing imagery and contains a number of ready-to-use live feeds that provide dynamic, real-time information on weather and Earth observation data that you can use to provide content with your data or the data you intend to collect.

Learn by doing

Hands-on learning through Learn ArcGIS will strengthen your understanding of GIS and of how you can use it to improve conservation land management and landscape conservation. Learn ArcGIS is a collection of free story-driven lessons that allows you to experience the application of GIS to real-life problems. The collection includes these and other lessons applicable to conservation:

- **Introduction to ArcGIS Online:** Get started with web mapping using ArcGIS Online.

- **Manage data in ArcGIS Online:** Provide data to city employees managing interactions between citizens and wildlife.

- **Integrate maps, apps, and scenes to tell a story:** Share information about earthquake risk using maps, apps, and scenes.

- **Propose wildlife corridors for key species:** Identify ideal habitats and connect them with a least-cost path.

- **Predict deforestation in the Amazon rain forest:** Map the impact of roads on deforestation.

- **Monitor whales with a multilingual survey:** Create a survey to record sightings of humpback whales near Costa Rica.

- **Build a model to connect mountain lion habitat:** Find suitable corridors to connect dwindling mountain lion populations.

- **Identify an ecological niche for African buffalo:** Determine ideal habitats using the R-ArcGIS Bridge.

Learn how to track illegal activity, monitor wildlife populations, manage wildlife conflicts, and more. Try these and other Learn ArcGIS lessons for conservation at learn.arcgis.com.

Get there faster with GIS templates

ArcGIS Solutions contains preconfigured app collections available in ArcGIS Online that are designed to reduce the time needed to deploy location-based solutions in your organization. You can use geographic information and the investment your organization has made in GIS to manage wildlife populations, track activities that threaten protected areas, run infrastructure smoothly, and conduct community outreach in and around wildlife areas. The apps in this next list can also help you manage protected areas, monitor conservation easements, conduct photo surveys, and much more.

Manage protected areas

- **Conservation Outreach** can be used to track wildlife conflicts and engage stakeholders in and around protected areas.

- **Protection Operations** can be used to track poaching or illegal activity and monitor the status of protection operations in and around protected areas.

- **Wildlife Management** can be used to capture wildlife observations and monitor the status of wildlife populations in and around protected areas.

- **Park Infrastructure** can be used to streamline the creation, evaluation, and maintenance of park infrastructure assets.

Monitor conservation easements

- **Conservation Easement Monitoring** can be used to routinely inspect conservation easements and monitor conservation easement programs.

Conduct photo surveys

- **Wildlife Photo Survey** can be used to publish photos collected from camera traps and conduct surveys that may identify animals in their natural habitat.

Learn more

For additional resources and links to examples, visit the book web page:

go.esri.com/pop-resources

CONTRIBUTORS

Matt Ball
Jim Baumann
Chris Chiappinelli
Keith Mann
Monica Pratt
David Smetana
Citabria Stevens
Alexa Vlahakis
Carla Wheeler

ABOUT ESRI PRESS

AT ESRI PRESS, OUR MISSION IS TO INFORM, INSPIRE, AND teach professionals, students, educators, and the public about GIS by developing print and digital publications. Our goal is to increase the adoption of ArcGIS and to support the vision and brand of Esri. We strive to be the leader in publishing great GIS books, and we are dedicated to improving the work and lives of our global community of users, authors, and colleagues.

Acquisitions

Stacy Krieg
Claudia Naber
Alycia Tornetta
Craig Carpenter
Jenefer Shute

Editorial

Carolyn Schatz
Mark Henry
David Oberman

Production

Monica McGregor
Victoria Roberts

Marketing

Sasha Gallardo
Beth Bauler

Contributors

Christian Harder
Matt Artz
Keith Mann

Business

Catherine Ortiz
Jon Carter
Jason Childs

For information on Esri Press books and resources, visit our website at esri.com/en-us/esri-press.